Praise

Dollars to Doughnuts is a thorough, objective assessment of the Earth's sustainability crisis. But more importantly, it outlines a host of actions we, as individuals, communities and governments, should embark upon to try to bend the needle towards a sustainable future. It's a great read, not too technical and not superficial. It should be in anyone's library who seeks a better, sustainable world.

> *Robert D. Mangold, PhD*
> *Pacific Northwest Forest Research Station*
> *USDA Forest Service (retired)*

Dollars to Doughnuts impressively details climate change issues that confront the world today. At the same time, it is frightening, amazing, and disturbing that WE have not taken this issue seriously. Hopefully *Dollars to Doughnuts* will receive the attention it deserves and inspire real behavioral changes— time is running out.

> *Chuck Mack*
> *Co-Chairman*
> *WCTPT*

Dollars to Doughnuts compellingly communicates a hopeful vision for a sustainable future by integrating the myriad facets of this complex challenge through a visionary's kaleidoscope. Author Prentice not only illuminates the potentially catastrophic outcomes of non-action; he also offers practical ideas, resources, and strategies for sustainable living that each of us can adopt, learn from, and initiate. This book is relentlessly researched; poetically descriptive; and incredibly comprehensive in identifying the multiple facets comprising a green energy transition. It covers everything from the influence of population growth, values, and culture through technology, media and economics. As an accomplished novelist, Mr. Prentice injects humor, metaphors, everyday examples, personal anecdotes, and famous quotations that transform what can be a dense, academic topic into a living, breathing entity. If you are seeking to understand the urgency and complexities of our planet's growing unsustainability, this is the book to read.

Sandy Judd, M.S.
Certified Health Coach
Science & Health Writer

DOLLARS TO DOUGHNUTS

IMAGINING A SUSTAINABLE FUTURE

LLOYD R. PRENTICE

Writersglen
PUBLICATIONS

ISBN 978-0-9825892-7-4 (Paperback Edition)
ISBN 978-0-9825892-8-1 (ePub Edition)

Publisher's Cataloging-in-Publication Data

Names: Prentice, Lloyd R., author.
Title: Dollars to doughnuts : imagining a sustainable future / Lloyd R. Prentice.
Description: Includes bibliographical references and index. | Marshfield, MA: Marshfield, MA, 2026.
Identifiers: LCCN: 2025926281 | 978-0-982592-7-4 (print) | 978-0-9825892-8-1 (ebook)
Subjects: LCSH Climatic changes. | Climate change mitigation. | Sustainability. | Global warming. | Environmental protection–Citizen participation. | Nature–Effect of human beings on. | BISAC NATURE / Environmental Conservation & Protection | SCIENCE / Global Warming & Climate Change
Classification: LCC QC903.15 .P74 2026 | DDC 363.73874–dc23

LCCN: 2025926281

To my grandchildren, Brooke and Lily

And to Professor Kate Raworth, with gratitude for her inspiring work

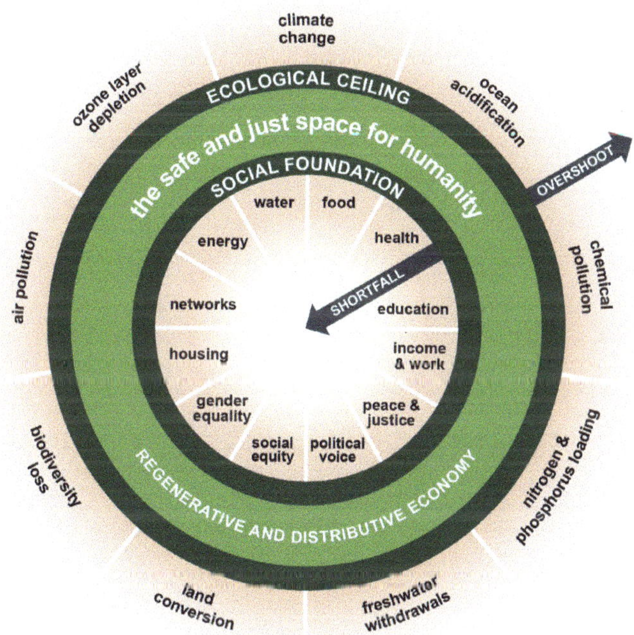

Figure 0.1: Doughnut Economics

Credit: Doughnut Economics Action Lab CC BY-SA 4.0

Acronyms

FAO	United Nations Food and Agriculture Organization
GDP	Gross Domestic Product
GGH	Gross Global Habitability
GGR	Gross Global Resources
GHW	Gross Human Welfare
GNH	Gross National Happiness
GNP	Gross National Product
IEA	International Energy Agency
IMF	International Monetary Fund
IPCC	Intergovernmental Panel on Climate Change
MIT	Massachusetts Institute of Technology
NATO	North Atlantic Treaty Organization
SEEA CF	System of Environmental Economic Accounting Central Framework
SEEA EA	System of Environmental Economic Accounting Ecosystem Accounting
TFR	Total Fertility Rate
UN	United Nations
UNESCO	United Nations Educational, Scientific, and Cultural Organization
US	United States of America
USD	United States Dollar
WBG	World Bank Group
WHO	World Health Organization

Contents

Introduction

This book is about agency— our power to act. Or not.

Imagine... you've hiked up treacherous trails over arduous mountain passes to a remote wilderness cabin. First night— winds howl. Cabin shakes. Then, dead silence. You awake to find the water pump frozen. You open the cabin door to a solid wall of icy white. You have firewood. Can melt snow. You light a fire in the massive stone fireplace but the chimney is blocked with hard-packed snow. Pine-scented smoke pollutes the cold cabin air. You cough, shiver. Fire burns low. Temperature drops. Not a soul within 30 miles to dig you out. What to do?

What would you do?

Or... your interstellar spacecraft is off course. Life-support systems flashing red. Navigation systems beyond repair. Your crew is rebellious— one faction storms the weapons lockers. Others look to you to bring the ship under control. What to do?

What would you do?

Or... think back. Remember that iconic Apollo 8 photo of our fragile spaceship Earth rising over the moon's horizon?

We know now that the life-support systems of our beautiful planet are flashing red. Temperature rising. Wars raging. What to do?

Temperature rising?

Oh, you know the story. Greenhouse gases accumulate in the atmosphere and trap solar heat reflected off the surface of the Earth— greenhouse effect.

Figure 0.2: Earthrise
Credit: NASA

Greenhouse effect?

If you've ever opened the door of a car parked long in hot sun with windows rolled up, you know most of what you need to know about greenhouse effect.

Globally, 2024 was the warmest year on record.

Since the 1850s global average temperature has risen at least 1.28 °C (2.3 °F).[1]

The Arctic's northern Barents Sea is a hot spot, heating up seven times faster than lower latitudes, with unknown consequences for the Barents Sea ecosystem.[2] Over the past quarter of a century, temperatures across the Arctic as a whole have risen four times faster than lower latitudes.[3]

Why?

Ice reflects more solar energy than land or open ocean. As sea ice melts, it exposes darker areas of land and water. These areas retain more solar energy and heat up more than reflective ice.

The geopolitical and climate implications of rising temperatures across the Arctic are significant. Open sea lanes and exposed land masses invite fossil fuel and mineral exploration and extraction. Russia, China, and NATO members Canada, Denmark, Finland, Iceland, Norway, Sweden, and the United States of America (US) contend for sovereignty and rights.[4]

Melting permafrost due to higher temperatures in the Arctic threatens infrastructure such as roads, railways, pipelines, and buildings supporting entire communities.[5] And more, melting permafrost releases greenhouse gases into the atmosphere. It may also release viruses and other dangerous pathogens.[6]

Oregon State University ecosystem scientist William J. Ripple and colleagues tell us that despite decades of warnings by climate scientists, extreme climatic conditions are now in uncharted territory. Escalating global temperatures caused by human activities have placed life on Earth under siege. And more, they say that "...time is up."[7]

And wars—

As I write, wars are raging around the globe. Ukraine. Gaza. Sudan. Myanmar. Elsewhere, tensions are high. Taiwan encircled by Chinese military forces. Vladimir Putin and Kim Jong Un threatening to launch nuclear weapons.

The 2024 Global Peace Index lists 92 countries involved in 56 international conflicts, the most since World War II:

> The rising number of minor conflicts increases the likelihood of more major conflicts in the future.[8]

By my reading of history, wars rate among the most deplorable manifestations of human behavior. Yet they persist. Pick a century, name a year, and it's likely that somewhere on Earth war was reaping human lives.[9]

Why?

I do understand that many profit from war and not a few seem to enjoy it. And certainly when a people are attacked by aggressive actors they have little choice but to defend themselves. Climate change is also a factor.

Human rights advocates report that drought, flooding, and extreme weather have driven conflict around the world and, at the same time...

> ...warfare has devastated ecosystems, imperiled access to vital resources and left behind toxic legacies that sicken civilian populations.[10]

Violent conflict over land and other resources dates back at least 10 millennia.[11] Yet devastating wars persist with little indication that they will be footnotes of history any time soon. For eight decades now we've lived under threat of nuclear annihilation.

The study of global warming extends back two centuries.

In 1824, French mathematician Joseph Fourier hypothesized that gases in the atmosphere must trap heat.[12]

In 1856, Eunice Foote, a scientist and women's rights campaigner from Seneca Falls, New York, confirmed the link between CO_2 and global warming with a simple backyard experiment involving four thermometers, two glass cylinders, and an air pump. She filled one cylinder with CO_2 and water vapor; the other with ordinary air. When she measured the gases in both sunlight and shade, she found that CO_2 and water vapor absorbed more heat than ordinary air. She concluded that more CO_2 in the atmosphere could warm the planet.[13]

A male colleague read her paper to scientific peers during the 1856 meeting of the American Association for the Advancement of Science. Women weren't allowed. Her findings were met with indifference by the scientific community.

In 1896, the Swedish physicist Svante Arrhenius concluded that "every time the CO_2 concentration in the atmosphere doubles, Earth's temperature will rise between 2 and 5 degrees Celsius (3.6 and 9 degrees Fahrenheit)."[14]

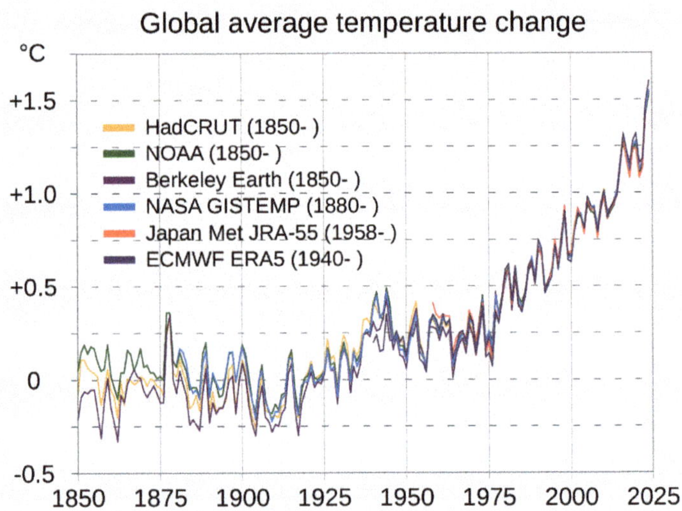

Figure 0.3: Correlation of measured global average temperature, from several different scientific organizations

Credit: RCraig09, CC BY-SA 4.0, via Wikimedia Commons[15]

We know now that rising temperatures around the globe are an existential issue for humanity. Environmental economist Giacomo Falchetta and colleagues estimate that the number of people worldwide at risk of chronic extreme heat exposure will likely double or more by 2050 due to global warming.[16] Indeed, global warming stresses nearly every ecosystem on Earth. We know the causes and how to bring them under control. Yet powerful economic interests and political elites choose cynical denial and obstruction over decisive action.

A 1968 report prepared by Stanford Research Institute scientists E. Robinson and R.C. Robbins for the American Petroleum Institute warned that rising CO_2 levels in the atmosphere could lead to serious environmental damage worldwide.[17]

In 1977 Marathon Oil reported that global temperature rise linked to "industrial expansion" could one day cause "widespread starvation and other social and economic calamities."[18]

That same year scientists at ExxonMobil submitted reports to management that fossil fuels contributed to global warming.[19]

What did Exxon management do? Denied. Authorized millions of dollars to pervert public opinion and delay action. From 1998 through 2019, ExxonMobile spent more than $37 million USD to cast doubt on global warming.[20]

The fossil fuel industry — oil, gas, coal — piled on. A three-year investigation by the Democratic staff of the US House Committee on Oversight and Accountability confirmed that fossil fuel companies have known that fossil fuels contribute to climate change since the 1960s. But they chose to deny— orchestrated a decades-long campaign to deceive the public and investors.[21]

The Institute for Policy Studies, in collaboration with the Climate Accountability Institute, analyzed tax returns of charities promoting climate disinformation and denial and the private foundations and donor-advised funds that support them. They found that 137 climate disinformation organizations received charitable donations from 2020 to 2022. Based on their findings, the researchers estimate that the total amount spent on climate disinformation likely exceeds "$219 million into the billions of dollars."[22]

In 2015, in response to the growing urgency of climate impacts, nearly every country in the world signed onto the Paris Agreement, a landmark international treaty under which 195 nations pledged to hold the Earth's temperature to "well below 2° Celsius above pre-industrial levels," and, going further, aim to "limit the temperature increase to 1.5° Celsius above pre-industrial levels."[23]

Oh, did I mention? BBC reported in February 2024 that "For the first time, global warming has exceeded 1.5°C across an entire year, according to the EU's climate service."[24]

What to do?

What can we do?

This book is not about climate change, nor war. It's about imagining a sustainable post-climate-crisis future. But climate change, environmental assault, and war provide context. It's taken our kind three centuries to stumble into impending planetary catastrophe. When and how did we go wrong? Without understanding missteps it's near hopeless to imagine a sustainable, just, and life-affirming future.

The jury is out, but hopeful signs nudge optimism that humanity will curtail greenhouse gas emissions before all terrestrial life on Earth dies of heat death.

— Independent energy think tank Ember reports that wind and solar are the "fastest-growing electricity sources in history."[25]

— Over the decade from 2009 to 2019 the price of electricity from solar declined by 89 percent.[26]

— Battery costs plunged from $1,200 USD per kilowatt hour (kWh) in 2015 to $150 USD per kWh in 2023 and, by one estimate, are likely to continue their downward trajectory toward $32 to $54 USD per kWh by 2030.[27]

Ending war is an iffier matter. ICAN, the International Campaign to Abolish Nuclear Weapons, reports that "...in 2023, the nuclear-armed states spent $91,393,404,739 USD on their arsenals — equivalent to $2,898 USD a second," up 13 percent over the previous year.[28]

The Bulletin of the Atomic Scientists' Doomsday Clock as of April 2025 is 89 seconds to midnight— "...a moment of historic danger."[29]

What can we do?

Kate Raworth's *Doughnut Economics: Seven Ways to Think Like a 21st-Century Economist* provides a powerful thematic metaphor to guide our thinking about sustainability. Her book diagrams the safe and just space for humanity as a doughnut. The doughnut vividly illustrates the boundaries between the outer ecological ceiling of resource availability and the inner social foundation of basic human needs.[30]

For humanity to thrive in a resource-constrained world, Raworth tells us, we must fundamentally revise economic assumptions and practices. We must move away from today's economic-growth-at-all-cost drivers of natural resource extraction, transformation, consumption, and waste toward circular energy and material flows that function within the reproductive capacity of nature.

Kate Raworth's book is consciousness-altering— an operating system for navigating spaceship Earth.

Inspired by Professor Raworth's brilliant metaphor, this book asks, "How can we best traverse the thorny maze from here-and-now to sustainability?" It's a reflective meditation on the opportunities and the challenges— what will it take to assure well-being and abundance for our children and all who follow in a natural resource-constrained world? This book is about limits and opportunities. Culture and economics. Governance and the power of individuals to influence change.

Sustainability... habitability of Earth... the welfare of future generations is in our hands. But what to do?

What can we do?

Deny? Delay? Party on? Let the tree huggers worry? Trust in higher powers?

Or, roll up our sleeves and do what we can?

My take? There's only one answer. We must act.

As individuals we have little agency in the face of global issues. We can, however, acquaint ourselves with the facts and consider contending ideas, points of view, and alternative courses of action. We can advocate. Agitate. Innovate. The clock is ticking. We must do all we can.

The goal of this book is to stimulate thought, research, discussion, and, dare I hope, constructive action. The path to sustainability is the work of millions. Here I pose questions that may help illuminate that path— naive, aggravating, but essential questions.

Climate science and sustainability research are fast-evolving topics. By the time you read this, many details will be yesterday's news. I'm not a climate scientist, policy specialist, nor expert in any topic covered. But, as an editor advised me long ago, "You can become expert on any topic by asking the right questions and doggedly tracking down answers from the real experts in the field."

Think of me as a doggedly curious writer and grandfather with deep concerns about the future and well-being of my children and grandchildren— and yours.

Two pressing questions inspire this book:

1. What kind of world do we want for our children and theirs?
2. How must we manage affairs on spaceship Earth to achieve it?

This book is about urgency. If we don't act— what then?

Part I

The Challenge

The best way to predict the future is to create it.

PETER DRUCKER 1991

1

Population

June 29, 2024

Yow!

I'm sitting here watching numbers flip by. Going up and up.

Current World Population

https://www.worldometers.info/world-population/

World population now 8,118,759,299. Oops! 8,118,760,320, no, no... up, up as I watch, fast approaching 8,119,000,000 and on up. Check it out. No doubt the number will be higher when you do. Within the last six months world population up more than the total population of Saudi Arabia. Thirty six million more mouths to feed within just six months; 36 million more young brothers and sisters of the world with lives before them; hopes and dreams to fulfill.

World population has quadrupled within my lifetime.[31]

Now more than 8.1 billion people with so much in common— shared need for air, water, food, shelter, and community. But in our ability to meet basic needs we couldn't be more diverse.

The richest one percent of the world's adult population owns almost half of total global wealth.[33] Some 3,028 individuals are collectively worth $16.1 trillion USD, says *Forbes* magazine.[34] This is more than $5.3 billion each, with no practical limit to opportunity, appetite, or ability to satisfy wants.

Yet, one in 11 human beings globally is afflicted by hunger according to a 2024

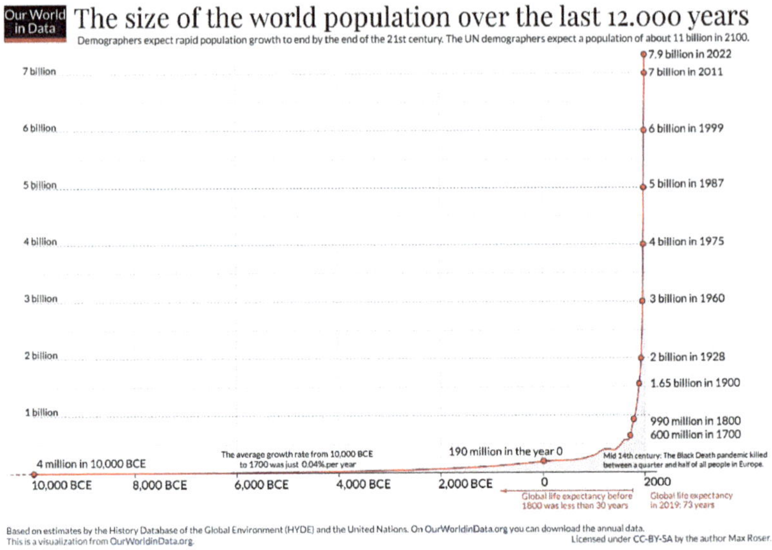

Figure 1.1: The size of the world population over the last 12,000 years
Credit: Max Roser - Our World in Data[32]

UN report.[35] More than 10 percent of people on Earth struggle to secure the most basic life-sustaining needs.

In 2024 the World Health Organization (WHO) estimated that 149 million children are too short for their age, 45 million suffer low weight, and 37 million are overweight, many due to poor nutrition.[36]

And shelter...

According to the United Nations' (UN) urban development agency, some 40 percent of the global population lacks adequate housing, secure land, and basic water and sanitation services. Some 300 million are homeless— "...lacking any form of stable shelter."[37]

In my country, the United States of America, with by far the largest Gross Domestic Product (GDP) in the world, 771,480 people experienced homelessness on a single night in 2024, including nearly 150,000 children.[38]

Unhoused populations are at increased risk of death due to disease, as well as other serious health issues.[39] As global temperatures rise, the unhoused are also the most vulnerable to heat-related illness and death. A study by Boston University public health researcher Zihan Lin and colleagues found that between the years 2015 and 2022, 8,648 deaths occurred among people experiencing homelessness in Los Angeles County, California— approximately three per day.[40] During a blistering heat dome that descended over California's Los Angeles County in August 2023, nearly half of deaths from heat illness or heat exposure were among the unhoused population.[41]

Just this week, as I write, the US Supreme Court ruled that cities can enforce bans on homeless people sleeping outside in public places.[42]

Across the country, more than 770,000 people are in shelters or on the street. They're forbidden to sleep on sidewalks or in public parks. Hassled. Assaulted. Arrested. Where can they go? Some politicians and pundits are advocating internment camps for homeless people.[43]

In the US, natural disasters, the cost of housing, and more people seeking asylum have driven up the homeless population in recent years.

In some countries the cause of homelessness is far more dire. More than 6.5 million Ukrainians have been driven out of their country and another 3.7 million internally displaced by Russian missiles, bombs, and on-the-ground military aggression.[44] Nearly 2 million Palestinians have been displaced in Gaza by the Israeli/Palestinian conflict.[45]

Sudan has the largest internally displaced population in the world— 11 million people.[46]

Al Jazeera reports that, post-coup, the Myanmar junta has driven at least 3 million out of their homes (not to mention killing more than 5,000 civilians and taking more than 20,000 political prisoners).[47]

As conflicts in Ukraine, Gaza, Sudan, and Myanmar rage on, yet more are at risk of homelessness.

People are also driven from their homes by climate-driven sea level rise, floods, and drought. The World Bank projects that by 2050 as many as 216 million people in the world could be forced to move within their own countries due to slow-onset climate change impacts.[48]

Is the Earth now so resource poor that it can't provide sufficient nutrition nor shelter for all of humanity?

As to nutrition, the food is there. But current political, economic, and wealth inequalities make its equitable distribution "...debatable."[49]

As to shelter, it's a "...serious concern."[50]

The need for food and shelter, of course, is driven by population.

Total Fertility Rate (TFR) of a population is the average number of children born to a woman over her lifetime. *The Lancet* tells us that global TFR fell from 4.84 to 2.23 over the period from 1950 to 2021.[51] TFR above 2.1 leads to an increase in population size. When TFR is less than 2.1, population declines. The authors project that fertility rates will decline to 1.83 by 2050 and 1.59 by 2100.

Birthrate, that is, the number of live births per 1,000 people, is another measure of population growth or decline. Birthrates vary widely by country. African countries such as Angola, Niger, Mali, Chad, and Uganda have more than 40 live births per 1,000 people; the US has 11; many European and Asian countries have fewer than 10.[52]

Population dynamics suggest that humanity is between a rock and a hard place. Population growth stresses global ecosystems; declining populations provide fewer young people to support the elderly, which, in turn, creates social and economic stress.

Not all population forecasts foresee decline of total fertility rate. The website *database.earth* estimates that within one generation world population will exceed 9.4 billion— more than 10 billion within two generations.[53]

What then?

2

Limits

"Chill! Chill..." I hear people say. "Doom-mongers have worried about food and population since 1798, when economist Thomas Robert Malthus argued that rising population would lead to mass starvation.[54]

"Our kind are a clever lot," they say. "We learn. We innovate. We'll never run out of food since food shortages will drive up prices and higher prices will bring new technologies and innovators into the market. Economic growth will solve the problem of hunger. Unfettered markets drive economic growth.

"Cut spending," they prescribe. "Cut taxes. Austerity spurs economic growth."

You'll recognize these ideas as neoliberal ideology that has dominated economies worldwide since the early 1980s.[55]

This doctrine may harbor smidgens of truth. I, personally, have never suffered insufficiency of nutritious food. Eight well-stocked supermarkets and scores of restaurants are within easy drive from my home. Indeed, the gift of fossil fuels widely distributed through liberal markets has brought historic abundance to multitudes around the world.

But, turns out, not without downside— greenhouse gas emissions drive destructive climate events.

There's a lesson here.

Consider:

July and August 2022: Floods in Pakistan submerged an area roughly the size of Italy[56] and disrupted the lives of 33 million people. Rising waters displaced 8 million people and took 1739 lives, including 647 children.[57] The floods

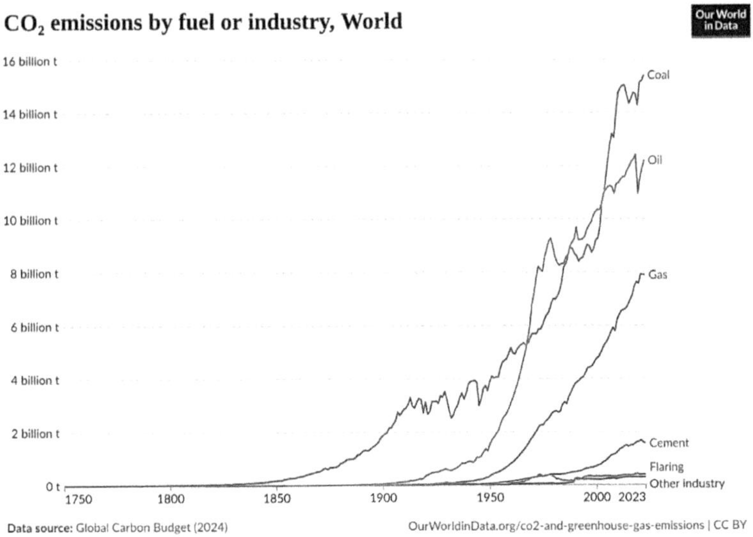

Figure 2.1: CO_2 emissions by fuel or industry
Credit: ourworldindata.org CC BY-SA 4.0[32]

demolished 897,014 houses,[58] destroyed or damaged 30,000 schools, and 2,000 health facilities.[59] More than 5.4 million people were forced to rely solely on contaminated water from ponds and wells.[60] Flood damages and economic loss exceeded $30 billion USD; reconstruction costs likely to exceed an additional $16 billion USD.[59]

March 2023: Canadian wildfires burned boreal forests equivalent to half the land area of Norway.[61] The Canadian wildfires released 647 megatons of carbon, more than seven of the top 10 carbon-emitting countries in 2022, including Germany, Japan, and Russia.[62]

July 23, 2023: Maricopa County, Arizona, was so hot that people were brought into the emergency room with significant, sometimes life-threatening, burns. People literally burned to death by falling onto hot asphalt paving.[63]

August 2023: 50,000 migrants took the week-long trek through the dense jungle connecting Columbia and Panama, overwhelming the capacity of medical treatment posts on the Panamanian border. Many were motivated by drought and crop failure.[64]

2023: Earth had the hottest three-month period on record, with unprecedented sea surface temperatures and much extreme weather.[65] Lakes and rivers rapidly warming, losing oxygen; aquatic life at risk.[66] The US set a record for billion-dollar disasters in a single year.[67]

Extreme weather disasters are driven mainly by high levels of CO_2 in the atmosphere. CO_2 concentration is largely the result of fossil fuel combustion. Since 1960 global atmospheric levels of CO_2 have increased nearly 33 percent.[68]

We confront limits and the consequences of defying them in every aspect of our lives. Extreme weather is but one example of vast harm inflicted by defying limits. Here are others:

— Material consumption beyond limits challenges the regenerative resiliency of natural systems.

— By-products and waste from production and consumption further degrade natural systems.

— Careless misuse of technology leads to unanticipated harm.

Figure 2.2: One small step for mankind
Credit: Joãao Gabriel

Examples:

— Nitrogen runoff from over-fertilized farms poisons waterways.[69]

— Overfishing, logging, and toxic pollution contribute to species extinction.[70]

— Combustion of fossil fuels contributes to global warming.

— Excessive use of antibiotics in factory farms contributes to evolution of "nightmare bacteria" that can cause fatal infections.[71]

Implication: disruption of natural systems imposes limits upon economic growth.

Questions:

— How accurately can we measure and monitor environmental limits?

— Have we approached or surpassed critical environnmental limits?

— If so, what then?

In 1970 the Club of Rome[72] commissioned Massachusetts Institute of Technology (MIT) system dynamics researchers to study exponential economic and population growth given finite supply of resources. Donella H. Meadows and colleagues developed a dynamic systems model of population and Earth resources. They ran widely varying assumptions through the model, postulating various rates of five factors that interact to limit global economic growth:

— Population increase

— Agricultural production

— Nonrenewable resource depletion

— Industrial output

— Pollution generation

In their 1972 book, *Limits to Growth*, they concluded that Earth's natural resources "...probably cannot support present rates of economic and population growth much beyond the year 2100." But the authors also offered a ray of hope: "The challenge of overshoot from decision delay is real, but easily solvable if human society decided to act..."[73]

Millions of copies of *Limits to Growth* were purchased worldwide. The book incited vociferous criticism, but subsequent studies with updated resource estimates have confirmed the harsher findings.[74]

Global population has more than doubled since 1970; consumption of natural resources is stressing critical ecosystems around the world; and extreme weather events are ever more costly and deadly.

Economist Herman Daly challenged the doctrine of economic growth without limits during his tenure as senior economist at the World Bank from 1988 to 1994 and, earlier, in his 1973 anthology *Toward a Steady-State Economy*.[75]

Economist Kate Raworth has proposed a framework for steady-state economics in her 2017 book *Doughnut Economics: Seven Ways to Think Like a 21st-Century Economist*.[76]

So now we confront key concerns:

1. Can a steady-state, natural-resource-constrained economy provide well-being, opportunity, and happiness for all of humanity?
2. Can we thrive while respecting and protecting the regenerative capacity of nature?
3. Can we transform our worldwide cultures, economic systems, and governance in time to avoid environmental catastrophe?
4. If so, how?

What to do?

What can we do?

3

Resources

Humanity on Earth exists by grace of intricately intertwined natural systems. We've long assumed that we can take fossil fuels from the Earth, fish from the sea, trees from the forests, crops from the soil, and water from rivers, lakes, and aquifers without limit or consequence.

We've presumed that we can dominate nature.

We've clear-cut forests; cultivated prairies and savannas; monocropped grain and maize season after season; poisoned soil with excessive fertilizers and pesticides; built urban sprawl on arable land, swamps, and wetlands; pumped water from ancient aquifers; gillnetted fish; trawled ocean floors; leveled mountains for minerals and fuel; discarded toxic waste into streams and landfills; spewed radioactive particles into the atmosphere; and enslaved fellow human beings— all in the name of economic growth and profit.

No doubt our creative intelligence and industry have brought the good life to many.

We've built iconic cities, subdued infectious diseases, split atoms, hacked DNA, probed distant realms of outer space.

For two centuries now we've been on a roll with dazzling visions and achievements yet to behold.

Growth.

But we know now that nature's bounty has limits. Stress one skein and the interwoven ecosystems of nature unravel. Our debt to nature is coming due. It's a hard lesson for many. Sunk investment, life-long careers, and vast fortunes are

Figure 3.1: Natural resource extraction
Credit: Homo Sapiens Foundation CC BY-SA 4.0[32]

at stake. Denial and resistance are palpably loud.[77]

Climate change is a hoax, we've heard. Wind and solar are not reliable. Transition from fossil fuels is too expensive. Solar panels and windmills are ugly. Windmills kill birds and whales.[78]

But ever more frequent and extreme billion dollar hurricanes and typhoons, extended droughts, wildfires, sea-level rise, and heat deaths in farm fields and major cities are powerful signals that we need to rethink long-cherished economic assumptions.[79]

We need to learn how to once again live within nature's grace. We need to redefine economic growth.

Kate Raworth's doughnut metaphor vividly illustrates the revised assumptions and real-world constraints we need to build into our culture, economic models, policies, and transactions. But given human nature and political realities, how can we possibly transition from growth-without-limits to doughnut economics within the few decades of opportunity dictated by climate change?

At best it's a race against time.

Case in point: Rensselaer Polytechnic Institute ecologist Kevin C. Rose and colleagues report that...

> Across all aquatic ecosystems, from streams and rivers, lakes, reservoirs, and ponds to estuaries, coasts, and the open ocean, dissolved oxygen (DO) concentrations have rapidly and substantially declined in recent decades.[80]

Lakes and reservoirs have lost as much as 5.5 to 18.6 percent of dissolved oxygen since 1980 as consequence of greenhouse gas emissions. Oceans have lost some 2 percent of dissolved oxygen since 1960 with much variability across marine ecosystems.

How long before we see hypoxia-driven mass die-off of aquatic life?

Fish and zooplankton need oxygen to survive. The concentration of dissolved oxygen in the ocean is inversely related to temperature. As temperature goes up, the concentration of dissolved oxygen goes down. Oceanographer Richard J. Matear and colleagues attribute the decline of dissolved oxygen in the oceans to global warming.[81]

The Intergovernmental Panel on Climate Change (IPCC) has warned of critical environmental thresholds that, if exceeded, will lead to irreversible consequences:

> ...rapid changes to the Earth system undermine critical life-support systems, with significant societal impacts already felt, and they could lead to triggering tipping points that irreversibly destabilize the Earth system.[82]

In 2023 an international team of 51 scientists working under the auspices of the Earth Commission published their study of the stability and resilience of the Earth's systems and human well-being.

Conclusion:

> Humanity has exceeded safe limits for four of five planetary systems: climate change, biodiversity loss, land use change, fresh water use, and nitrogen and phosphorus cycles.[83]

The Potsdam Institute for Climate Impact Research identifies nine planetary boundaries:[84]

1. Climate change

2. Biosphere
3. Land system change
4. Freshwater change
5. Biogeochemical flows
6. Ocean acidification
7. Atmospheric aerosol loading
8. Ozone layer depletion
9. Introduction of novel entities, such as synthetic chemicals, to the environment

They report that we've exceeded seven of them.[84]

What to do?

What can we do?

And how?

The good news is that many transnational institutions, governments, nonprofits, and fired-up activists are on the case. Initiatives are actively pursued globally. See Appendix C. The goal across the board is sustainability of natural resources essential to human well-being— habitability of Earth.

At stake:

- The air we breathe
- The water we drink
- Food that nourishes us
- Raw materials for manufacture and construction
- The global envelope of life that sustains us

The prize: opportunity and abundance for all.

Opportunity?

Change implies leaving behind the old and embracing the new. We each have opportunity to define change for the better. If natural limits impose constraints on material consumption, perhaps we can cut back consumption and find satisfaction, gratification, and reward in cultural and artistic endeavor, public service, enriched human relations, and spiritual growth. Perhaps we can redefine the nature of work. Rather than labor long hours for ever more enrichment and glorification of billionaires, perhaps we can work shorter hours with dignity and pride at jobs that truly matter for our families and communities. If war diminishes natural resources and capacity to regenerate, why squander resources on war?

Abundance?

Here we flail in treacherous waters. Technicalities and subtleties long debated by political philosophers and economists lurk like sharks in these soundless depths. But swim we must since we're at the heart of our concerns— the capacity of natural systems to satisfy the basic needs of people worldwide.

Abundance— Merriam-Webster provides three definitions:[85]

1. An ample quantity
2. Affluence: wealth
3. Relative degree of plentifulness

The first definition suits us well. But, what's enough? How much is too much? Who gets how much of what? And who decides?

"No problem," economists have argued, "It's all a matter of how much people are willing to pay." But what price would we put on the last stand of old growth redwood trees or the last codfish bottom-trawled from a fished-out sea?

And price?

What's a fair price for a commodity, product, or service? Denominated how? Cost plus? Whatever the market will bear? Some companies are setting prices based on their customers' personal profiles gleaned from credit card and Internet history.[86]

What's a fair price for basic human needs such as food and habitable shelter? If food is scarce should the wealthy feast while the poor starve? Our world today is blighted by appalling inequalities and waste.[87]

The World Food Programme tells us that, "Around one fifth of food produced for human consumption is either lost or wasted, amounting to a financial loss of about $1 trillion USD annually."[88]

In a natural-resource-constrained world, what's a fair price for resource-intensive wants? The human appetite for "wants" is boundless. Wolfgang Hoeschele explores how wants are "...consciously generated in order to create profitable scarcities."[89]

Profitable scarcities?

To note a few:

— Think high food prices driven by oligopolistic supply chains.[90]

— Think of massive mark-ups of prescription drug prices by pharmacy benefit managers.[91]

— Think name-brand prescription sunglasses marked up 1,000 percent.[92]

— Think advertising that appeals to basic drives and insecurities for everything from automobiles to skin creams.[93]

To the point, not all scarcities are imposed by natural resource limits. Many scarcities are artificially imposed in pursuit of profit. Some are even enforced by law— think patents and copyrights.

And so, we circle back. Who gets how much of what? And who decides?

Given the reproductive capacity of natural resources, can we find equitable balance between the pursuit of profit, our collective appetite for natural resource-intensive wants, and the universal need for food, water, and shelter?

The capacity to sustainably harvest natural resources to fulfill basic human needs comes down to metrics and management. But management by whom and how? Turns out that many institutions and individuals are developing, promoting, and implementing management strategies to assure sustainability of natural resources:

— fisheries[94]

— forests[95]

— agriculture[96]

— manufacturing[97]

— recycling[98]

— materials[99]

— ecosystems[100]

- mining[101]

But the challenges are formidable—fear of change, vested interests, competing visions, political conflict, exploitative political and economic autocracies, power rivalries, misinformation and disinformation.[102] What to do?

What can we do?

4

Metrics

How much is too much?

Record-setting extreme weather events driven by CO_2 concentration in the atmosphere are urgent signals that activities of mankind are approaching existential limits.

Species extinction is another— estimated rates of extinction among bacteria, fungi, plants, mammals, birds, reptiles, amphibians, fish, and invertebrates are between 1,000 and 10,000 times higher than natural rate of species loss.[103]

But what if such after-the-fact signals are coming to our attention too late? Consider— a speeding ticket is a rather inconvenient way to monitor highway speed. A speedometer is considerably more appropriate.

We need existential threat alerts in ample time to take effective action.

GDP is a widely accepted metric of prosperity with significant influence on economic policy. But is it the best indicator in this era of global warming?

Think of economies around the world as cars and trucks on a treacherous mountain highway; GDP as their dashboard speedometers.

"Faster, faster..." pundits and politicians urge over the radio. "Drive faster, ever faster and you'll get there sooner."

But if road conditions are icy or foggy will they get there safely? Or even get there at all? And, more to the point, get where?

"Economic growth is good," say the pundits. "Ever rising GDP brings prosperity to all." This story makes intuitive good sense. We're proud to see our children grow from infancy to adulthood. But consider how dismayed we'd be if their childhood growth rate continued through their 20s, 30s, and 40s. Consider growth of cancer cells, metastasis.

GDP is one measure of a country's economic health. It includes consumer and government spending, net exports, and total investments within a one year timeframe.[104] It omits, however, externalities such as evironmental and public health costs imposed by overexploitation of natural resources and careless disposal of the by-products and waste of human production. Think failed fisheries,[105] unsustainable water practices,[106] greenhouse gases, "forever chemicals" such as PFAS,[107] and microplastics.[108]

GDP is a poor measure of human well-being in other realms as well. Consider a year of rising crime, drug addiction, military build-up, or extraction of natural resources beyond reproductive capacity. Money spent on such events inflates GDP but tells us little about prosperity. Nor does GDP tell us how goods and services are distributed across the population. GDP metrics include the cost of reconstruction following a devastating tornado but not the stress, trauma, and disruption of human lives nor loss of treasured family heirlooms. Neither does GDP measure the toll on human well-being, such as homelessness, child hunger, or health outcomes.[109]

The WHO reports that one in seven 10-19-year-olds globally experience a mental disorder and that suicide is the third leading cause of death among those aged 15-29 years old.[110] How can we put a dollar value on teens suffering anxiety, depression, and hopelessness? On a 14-year-old daughter's suicide?

GDP does include spending on illegal drugs, prostitution, and illegal gambling, which, according to economist Timothy Taylor, is a good reason why it should not be treated as a measure of social well-being:

> ...social welfare would clearly be improved if crime was lower and money spent on discouraging and punishing crime could instead flow to something that provides positive pleasures and benefits.[111]

Indeed, given growing disparities of income and wealth in the US, the benefits of growth clearly do not trickle down to lower quintile families and individuals— yet another shortcoming of GDP as a policy-driving metric.

In short, GDP does not provide actionable insight into over-exploitation of critical natural limits nor adverse social trends.

Without sensitive and timely environmental and social metrics we're wearing cognitive blinders while attempting to manage our economies. Indeed, we might as well be blind.

In 1972 His Majesty Jigme Singye Wangchuck, the Fourth King of Bhutan, advocated a metric that he called Gross National Happiness (GNH).[112] When Bhutan embraced democracy in 2008, GNH was written into its constitution: "The State shall strive to promote those conditions that will enable the pursuit of GNH."

The four pillars of Bhutan's GNH are:[113]

1. Good governance
2. Sustainable socioeconomic-economic development
3. Cultural preservation
4. Environmental conservation

The pillars have been further classified into nine domains:

1. Psychological well-being
2. Health
3. Education
4. Time use
5. Cultural diversity and resilience
6. Good governance
7. Community vitality
8. Ecological diversity and resilience
9. Living standards

Gross National Happiness US, a tax-exempt nonprofit, is striving to replace GDP as a prime metric of US social and economic performance with GNH.[114]

The World Happiness Report, a partnership among Gallup, the Oxford Wellbeing Research Centre, and the UN Sustainable Development Solutions Network, was first published in 2012.[115] One or another Scandinavian country has always held top place. In 2025 Finland had the honor, with a score of 7.736 out of 10.[116]

Since 2020, the report has considered how environmental quality shapes how we feel and think about our lives.[117] To understand and respond constructively to environmental stresses that affect our happiness and health, however, we need metrics that are both specific and compelling.

Environmental sustainability metrics need to explicitly signal the state of essential stocks drawn down from nature, e.g., mineral resources, fisheries, forests, arable land, and more. They must measure the loss of essential resources due to human activities and interventions. They must be comprehensive, timely, conveyed as simply and dramatically as the Bulletin of the Atomic Scientists' Doomsday Clock, and reported as widely as GDP.

But how can we possibly measure the number of fish in the sea, trees in the forests, birds, bees, and plots of arable land? And how can we do it at planetary scale?

Vaclav Smil devotes a book to the topic— *Harvesting the Biosphere: What We Have Taken from Nature.*[118] His extensive review of satellite data and field research is technical and, for this reader, challenging, with measures of biomass (total mass of living matter within a given area), phytomass (plant biomass), and the mass of living animals measured in metrics such as grams per square meter or metric tons per hectare. As one can imagine, there is considerable uncertainty in many of the estimates.

Will metrics reported in Smil's book provoke congressional action? Fire up a mass movement of sustainability activists? Doubtful. We need brilliant storytellers for that. But the work of inventorying natural resources is well underway.

No doubt many sustainability metrics are defined, applied, and reported by national and international agencies. Some pop up as headlines— a few, such as concentration of CO_2 in the atmosphere, more prominently than others. It's not clear, however, how effectively they drive policy priorities.

Sad fact: economic, social, and sustainability metrics are often challenged and denied by vested interests.[119]

Questions:

— How can we improve and better report sustainability metrics?

— Why do we allow vested interests to put the future of humanity at risk by suppressing valid scientific metrics and denying implications?

5

Money

Peel back technical jargon and we find that GDP is all about money— money exchanged for tangible goods and services, including money paid out to rent money, that is, interest on debt.

Economists tell us that money has three functions:[120]

1. Store of value— think money put away for a rainy day
2. Unit of account— think your bank statement
3. Medium of exchange— think grocery shopping

Herman Daly and John B. Cobb Jr. tell us that it's not so simple:

> Money ranks with the wheel and fire among ancient inventions without which the modern world could not have come into being, but it is much more mysterious. It is a unit of account that changes size like a rubber yardstick, a store of value that can swell or shrink over time, a medium of exchange that often never leaves the bank, an interest bearing debt and non-interest-bearing debt, a commodity like gold and a non-commodity token (paper money). Easily transferable into real assets by individuals, it is not at all transferable into real assets by the community. Counterfeiters are sent to jail for making it, but the private banking system can create it out of nothing and lend it at interest. It is illegal for citizens to destroy or deface it, but private banks can annihilate it. Some economists think it merely a veil behind which real factors

Figure 5.1: Currencies
Credit: Wazouille

> determine economic life, others consider it among the most important of determinants; some think its quantity should be determined by a fixed rule, others that it should be manipulated by public authorities. And in addition, some people even claim that it inspires a love that is the root of all evil.[121]

Liza Minnelli sang a delightful lyric in *Cabaret*—[122]

Money makes the world go around

The world go around

The world go around

Taken literally, Liza Minnelli is wrong. The Earth rotates most handily without money. But in another sense she is dead on. Money is a token of admission to society. In urban societies one must have money to legally obtain the necessities of life.

One can also say that money is a denominator of trust. If I sell my well-used clunker for three crisp $100 bills, I trust that Staples will accept those bills for the HP printer I need.

Excessive debt, on the other hand, curtails full participation in society— limits options and agency. And some debts can't be repaid. We can't pay off nature with coin of the realm for the products and services she delivers or the harms we inflict. Try to give a hundred dollar bill to a tree and see what that gets you.

A common view on the street is that money is a zero sum commodity if you get more, there's less for me. That was true when currency was literally gold, another scarce but valued commodity, or a legally recognized claim against such valued commodities. But no longer. In modern economies, central banks balance employment and inflation by setting a price on money loaned to commercial and retail lending institutions. Banks, in turn, lend reserves to consumers and businesses. This adds to money in circulation. The money is spent, deposited, and loaned out again.[123]

In time of war or economic distress the money spigot is cranked up to fund the crisis. Case in point, the International Monetary Fund (IMF) reports that during the COVID-19 pandemic governments put forth $9 trillion to prop up national economies:

— $4.4 trillion as direct budget support

— $ 4.6 trillion as "...additional public support loans and equity injections, guarantees, and other quasi-fiscal operations"[124]

In short, money supply in modern economies is the product of human judgment and, arguably, political calculation. Indeed, the copious font of money in our modern economy is debt.

Money is like rain on a mountainside. Raindrops trickle into small streams that join into rivers that flow into lakes and on down to the sea. Rivers can be dammed to generate power. Money accumulates through savings, profits, criminality, and taxes. Sufficient accumulation confers power to determine what tangible goods get produced and what consumed.

Money is a metric of power. And more, money is morally neutral. It matters little whether it is accumulated through honest labor, invention, taxation, exploitation, greed, extortion, fraud, or theft— money can be exchanged for the necessities and good things of life. It facilitates fulfillment of wants without end which, in turn, feeds motivation to exploit nature. When we exploit nature in excess of natural limits, however, no one wins. Hurricanes, floods, wildfires, drought, crop failure, and sea-level rise harm all in their wake— rich and poor. Indeed, over-exploitation of nature imposes onerous debt on all of humanity. When that debt comes due, destruction, misery, and death follow.

This circles us back to Kate Raworth's doughnut metaphor. Sustainable economies function within limits illustrated by her metaphorical doughnut. Ample evidence has accumulated that, globally, we consume more than natural systems can sustain yet fail to provide basic needs for the neediest. Globally, a relative few have indulged and profited far more than others. This raises issues of accountability. Who, if anyone, should be held liable for global-scale environmental losses? At what point should we consider over-exploitation of natural resources a crime?

A record number of lawsuits globally are holding companies and governments to account for the ravages of climate change.[125] Conversely, the U.K. has been cracking down on climate protesters with harsh prison sentences.[126]

Questions:

— Should injury and material loss due to climate change be fairly compensated by greenhouse gas emitters?

— If so, how and by how much?

— Should exploitation of nature beyond regenerative limits be considered a crime?

— If so, how should such crimes be enforced and by whom?

— Should climate deniers be held to account?

— Should climate-motivated protest be considered a crime?

No doubt these questions will loom large over decades to come. Daly and Cobb put money up with fire and the wheel in shaping our modern world. But catastrophic climate-fueled wildfires in Australia, Canada, China, Greece, the US and elsewhere suggest that money invested in fossil fuel production has rained fire on the Earth. Money invested in alternative energy, mitigation of climate threats, and restoration of essential biosystems, on the other hand, will put us well along the path toward sustainability of human life on Earth.

Financial advisor Dave Ramsey tells us:

> ...money is not the key to prosperity. How you handle it is.[127]

Abenaki filmmaker Alanis Obomsawin puts it bluntly:

> When the last tree is cut, the last fish is caught, and the last river is polluted; when to breathe the air is sickening, you will realize, too late, that wealth is not in bank accounts and that you can't eat money.[128]

6

Value

I hold in my hand an apple. Quite reasonably you might ask, "How much did that apple cost?" or "What is the value of that apple?"

Figure 6.1: Apple picking

Credit: Tom Shockey Creative Commons Attribution 2.0[32]

I hesitate..."It depends."

So, picture: beautiful September Sunday afternoon. Sky bright blue. Rows of well-tended forest-green trees dappled with red and jade McIntosh apples. Lovers two-by-two, families with infants in strollers, scurrying preteens, young women in groups of three and four bob and weave among the rows of apple trees, filling white bags with their gleanings.

The white bags come in three sizes. Pickers must purchase one or more bags before they enter the orchard. My wife and I purchased the smallest bag for $16 USD. So I could divide $16 by the number of apples in the bag when we checked out to tell you how much my apple cost.

But that's not the whole story.

We drove 30 minutes each way to harvest this apple at cost of time, gas, and wear and tear on our car. When we returned home we stored the apple in the refrigerator. So cost of electricity is involved. Once I've consumed the apple, I'll toss the core in the garbage bin. We pay for garbage collection so we must consider cost of disposal.

Now, ask the accountant at the farm where I picked the apple how much it cost and he'll tell you an entirely different story.

If you're asking how much apples cost because you'd like an apple of your own, we could drive down to the supermarket. I'm sure prices are posted. We could even shop around and, perhaps, find cheaper prices.

Oh, I must have misheard. How much did this apple cost in Mexican pesos? Currency of exchange matters.

As to value...do you mean nutritional value? Value as a late afternoon snack? Or, perhaps, the value of the memory this apple evokes of a lovely afternoon shared with my loving wife?

Passing thought: I'd imagine that an apple would be priceless to the parent of a starving child.

Referring to nature, Berta Martín-López of the Social-Ecological Systems Institute at Leuphana University of Lüneburg in Germany defines three kinds of value:

1. Intrinsic values refer to the inherent worth of nature as an end in itself, regardless of any human interest.

2. Instrumental values represent the importance of nature as a means to achieve human ends or satisfy human needs, interests or preferences.

3. Relational values refer to those concerns related to the meaningfulness of relationships, such as those between nature and people and among people within nature or fostered by nature.[129]

I'd argue that there's more to the matter. Climate extremes and devastation of critical ecosystems threaten life on Earth. What's the value of nature at the outer limits of global heating and ecosystem collapse? Should it come to that, how much would we be willing to invest to pull back from the brink? Seems to me that we need a fourth definition of value that captures the significance and urgency of nature exploited beyond reproductive capacity.

Imagine that an evil alien from outer space has kidnapped you for ransom. Locked you in an air-tight room.

How much would you give to be released?

I conducted a not-so-random poll with sample-size one. Response?

"Everything."

Suppose the evil alien has locked a complete stranger in the air-tight room? How much would you give?

My guess— Opinion is divided.

So let's rephrase.

Suppose a message of unknown origin is broadcast simultaneously over every telecommunication channel on Earth—

"Earthlings, we have technology to completely obliterate all plant life on Earth and will do so unless you transfer all refined gold on Earth to us within one day counting down from NOW!"

As you know, plants produce oxygen that nearly all living animals, including humans, need to breathe. So what's the value of all plant life on Earth relative to the spot price of gold?

Our hypotheticals suggest that there is a fourth category of value— existential value to individuals, communities, and humanity as a whole.

I think now about the tree from which I plucked my apple. It took the tree a year to set and flesh out the apple I hold in my hand. It will take another year for the tree to replace it. This apple is filled with seeds— each with a probability of germinating and growing into another apple tree.

So, what's the value of this apple to the tree?

And this gets me wondering about trees in general.

How can we put a value on, say, old-growth forest in British Columbia, Canada? To the Wet'suwet'en First Nation in British Columbia the old-growth forest is a priceless heritage.[130]

To British Columbia Timber Supply (BCTS), a commercial arm of British Columbia's Ministry of Forests, they're worth some anticipated increment of the province's GDP.

Turns out that the United Nations (UN) has developed a statistical framework "...for organizing data about habitats and landscapes, measuring the ecosystem services, tracking changes in ecosystem assets, and linking this information to economic and other human activity."[131]

The system, called SEEA Ecosystem Accounting (SEEA EA), helps address questions regarding the relationship between the economy, society, and the environment and how we measure well-being and social progress.

- **Ecosystem Extent accounts** record the total area of each ecosystem, classified by type within a specified area.
- **Ecosystem Condition accounts** record the condition of ecosystem assets in terms of selected characteristics at specific points in time.
- **Ecosystem Services Flow accounts** (physical and monetary) record the supply of ecosystem services by ecosystem assets and the use of those services by economic units, including households.
- **Monetary Ecosystem Asset accounts** record information on stocks and changes in stocks (additions and reductions) of ecosystem assets. This includes accounting for ecosystem degradation and enhancement.

I doubt, however, that SEEA EA can put a value on a lovely day of apple picking with a loved one. And I wonder what value the Wet'suwet'en First Nation would place on this method of assessing the value of their ancestral land.

Clearly, when we consider the future of humanity, we must distinguish between value to "me" as an individual and value to humanity as a whole. And more, when we assign value, we must strive for balance between the interests of individuals, communities, and human society as a whole, and the natural systems that support all life on Earth.

Questions:

- How can we balance the interests of the individual and society, of humanity and the living Earth?
- How can we account for cultural traditions when ascribing value to natural resources?
- How can we assign value to services such as temperature modulation, water capture and storage, and pollution absorption?

7

Culture

Am I wrong?

As adults, we're each pretty much set in our ways. We share values, beliefs, language, and many daily rituals with family, friends, and important folks in our lives. Once settled into adulthood we follow more or less predictable career paths; associate largely with like-minded people.

Fundamentally altering our values, beliefs, and aspirations would be like clawing up a steep muddy slope, cultural gravity working against us every inch of the way.

It's in this sense that I use the word culture. We're embedded like fish in water; don't give it much thought. But culture influences the trajectories of our lives from cradle to grave. In this sense there are myriad and diverse cultures around the world many with drastically different values, beliefs, and social imperatives— cultural differences that all too often drive intercultural tension, conflict, and violence.

My notion of culture overlaps two concepts much studied by sociologists— in-groups and out-groups[132].

But I'm thinking more about the cultural influences of family, education, religion, career, and social participation on self-identity, moral compass, and social allegiance.

Note that my focus is on consumer cultures of the US and other so-called "developed" nations. Thriving within natural resource limits is likely to impose severe strains on the political and cultural identity and economic status of many.

Consider for a moment the formidable challenge of reconciling car culture in industrialized nations with long-term sustainable lifestyles. One can argue that cars are inextricably embedded in consumer cultures— everything from career choices and commute time on busy highways to acres of parking spaces on urban streets, city blocks devoted to parking garages, and sprawling parking lots in suburban shopping malls. And more, the auto industry — spanning manufacture through repair, highway maintenance, auto insurance, and fossil fuel supply chains — is a significant contributor to consumer economies.

Now, consider the environmental impact of cars— fossil fuel emissions, urban sprawl displacing fertile farmland, microplastics and other toxins spewed by the rolling tires[133] of some 1.5 billion cars and 278 million trucks traveling world highways.[134]

Fear of giving up the convenience and economic benefits of cars influences effective response to the climate crisis and rapid transition toward sustainable economies. How many are willing to sacrifice short-term comfort, convenience, and the economic contributions of the transnational car industry for long-term environmental gains? Among other things, cars are expressions of personal identity, as well as conspicuous status symbols. The politics of transforming car culture from fossil fuels to electrification has already inflamed in-group vs. out-group antagonisms.

What to do?

Resistance to proactive climate investments and policies is inevitable. Careers, life goals, ambitions, status — even identity itself — are at stake for countless people. For many, climate change is a toxic political issue. And let's not underestimate the paralytic power of fear of the new and the unknown. Question is, given climate projections, can we overcome culture-bound resistance to global sustainability initiatives? And can we do it in time? What to do?

What can we do?

Climate scientists and ecologists project alarming global climate and ecosystem scenarios that extend out to 2100 and beyond. Many fear that within our lifetimes, and well within our children's, we will exceed tipping points beyond which all bets are off with respect to habitability of vulnerable urban and agricultural areas globally.[135]

Global Tipping Point Project, collaborative research by more than 200 researchers and practitioners from over 25 institutions, reports that:

- Multiple drivers such as greenhouse gas emissions, deforestation, urbanization, and pollution are destabilizing interlinked functions of

Earth's natural systems.

- Some climate tipping systems closely interact, and most interactions tend toward destabilizing, making tipping 'cascades' possible.
- In near-term, we may be approaching several critical tipping points.[136]

If consensus climate projections are as accurate as ecoscientists claim, then we clearly need global coordination and cooperation to mobilize an all-hands-on-deck response.[135] Prudent response to climate threats will require rapid reorganization of energy, land, urban, and key infrastructure and industrial systems. But despite clear and present danger, cultural inertia and special interests stand in the way.

But consider— culture wars are provoked and amplified by special interests. Incentives to deny global warming and environmental plundering are blatant. As just one example:

In 2022, public financial support for fossil fuels, in the form of subsidies, investments by state-owned enterprises, lending from public financial institutions, and environmental costs, surged to $7 trillion USD globally— a record high.[137]

As another example: An analysis of the 118th US Congress found that 123 elected officials, including 100 representatives and 23 senators out of 535 total members, are climate deniers. Climate deniers in Congress wield significant influence on public perceptions of climate change as well as climate policy.

> The climate deniers in the current Congress have received $52,071,133 USD in lifetime campaign contributions from the fossil fuel industry.[138]

Corporate interests such as the fossil fuel, chemical, mining, and timber industries have legitimate concerns. The stakes are high. Billions of dollars in sunk investments are at risk. Extractive industries provide employment for millions, dividends and stock appreciation for pension funds, energy and materials for supply chains that weave throughout the economy.

So aggressive political efforts to influence public policy are understandable. Politicians, in turn, rise to power through support of self-interested constituencies. The political challenge, however, is the pressing need to negotiate a fair and just transition from fossil fuels and exploitive resource extraction to renewable energy and long-term sustainability.

Arguably, the fossil fuel industry has elevated self-interest over the interests of humanity. Cynical efforts to provoke culture wars, to manipulate cultural biases to suppress climate science and sabotage constructive policy for the sake of short-term profit and lifestyle convenience by leaders and influencers who know better is, well, I'll let you and children worldwide pass judgment.[139]

So, again: can we change deeply embedded cultural beliefs, values, aspirations, preferences, behavior, and institutions on global scale in sufficient time to avoid irreversible collapse of life-support systems on planet Earth?

The answer is a resounding "maybe." As you'll find in Appendix C, many around the globe agree on the urgency of the moment and are doing what they can to curtail fossil fuel emissions, regenerate degraded natural resources, and rethink cultural, lifestyle, and economic assumptions. And many are actively working on diverse fronts toward the goal of long-term global sustainability. The good news is that we now have effective technologies and more in R & D pipelines to reduce greenhouse gas emissions and revitalize degraded land. So change is in the air.

To walk the talk of sustainability, then, we must reevaluate our cultural bias toward consumption. We must see change as opportunity. To bring vested interests on board, we must craft incentives and policies to minimize losses and maximize gains of transition from the status quo to sustainability. To mobilize political will, we must counter fear mongering and denial, and support intellectual, corporate, cultural, and political leaders who demonstrate firm commitment toward a just transition from the here-and-now to sustainability. Above all, we must reconceptualize the concept of economic growth. Futurist Jesse Hirsh imagines a future in which money and markets are no longer the arbiters of political possibility. In this future, three ideas guide the collective stewardship of life:

- The Earth is finite.
- Human creativity is infinite.
- What we produce is bounded by material limits.[140]

Let's assume that we have the courage and will to embrace sustainability. Just what, then, do we mean— global sustainability?

In a nutshell, what are we working toward? What must be true in order to say, "Mission accomplished?" Clearly life as we know it in developed countries, for all its conveniences and pleasures, has led us off-track. But what are the alternatives? What might it feel like to live with different incentives, rewards, and economics?

The U.N. Department of Economic and Social Affairs 17 sustainability goals put the challenge of culture change in harsh perspective. As you review the list, imagine what must change and how:

- End poverty in all its forms everywhere
- End hunger, achieve food security and improved nutrition, and promote sustainable agriculture
- Ensure healthy lives and promote well-being for all at all ages
- Ensure inclusive and equitable quality education and promote lifelong learning opportunities for all
- Achieve gender equality and empower all women and girls
- Ensure availability and sustainable management of water and sanitation for all
- Ensure access to affordable, reliable, sustainable, and modern energy for all
- Promote sustained, inclusive, and sustainable economic growth, full and productive employment, and decent work for all
- Build resilient infrastructure, promote inclusive and sustainable industrialization, and foster innovation
- Reduce inequality within and among countries
- Make cities and human settlements inclusive, safe, resilient, and sustainable
- Ensure sustainable consumption and production patterns
- Take urgent action to combat climate change and its impacts
- Conserve and sustainably use the oceans, seas, and marine resources for sustainable development
- Protect, restore, and promote sustainable use of terrestrial ecosystems, sustainably manage forests, combat desertification, reverse land degradation, and halt biodiversity loss
- Promote peaceful and inclusive societies for sustainable development
- Strengthen the means of implementation and revitalize the Global Partnership for Sustainable Development[141]

The U.N. Sustainable Development Report 2024 rates progress of 167 countries toward these goals across some 125 indices, then ranks them on a 0-to-100 scale.[142]

Finland shows most progress with an 86.35 score; South Sudan shows least progress with a 40.14 score.[143]

Who can dispute these noble goals? A card-carrying neoliberal might argue that economic growth and market forces are sufficient to achieve many of them.[144]

But questions linger...

- Can projected global populations be "sustained" within limits of Earth's natural resources?
- Is sustainability even possible given human inclinations toward envy, greed, conspicuous consumption — not to mention prevailing inequities of power?
- What fundamental values, beliefs, and civic institutions must change and how to maintain globally sustainable cultures?
- How can we negotiate the many challenges of cultural adaptation and change within the time constraints dictated by climate projections?

Cultures are resistant to change. But climate change has, throughout history, devastated cultures around the world. Arizona State University professor Billie L. Turner tells us that two major droughts brought down the Mayan culture known for monumental architecture more than 1,000 years ago.[145] Global heating is now driving drought, terrestrial and marine heat waves, fire, floods, crop failure, political instability, and mass migration both within nation-states and across national borders— threatening thriving cultures around the world. Fortunately we have scientific early warning models and technologies to constructively transition from economic growth without limit to sustainability. But timely transition can't occur without consensus, political will, and urgent effort— that is, fundamental reassessment and transformation of consumer culture.

What to do?

What can we do?

Culture change requires charismatic leaders who inspire people in new directions, new stories that help change perception, and incentives that support new ways of thinking.

8

Incentives

July 8, 2024

Two stories caught my eye this morning as I browsed through various sources. The first was a series of in-depth interviews with homeless men and women in Russia.[146] The fact that the interviews were transacted in Russia is neither here nor there. No doubt many of the life circumstances recounted are experienced by homeless victims around the world.

Among other questions, interviewer Irina Kravtsova asked:

— "What's it like living on the street?"

— "What's keeping you from returning to 'normal' life?"

The stories are heartbreaking and bleak

Anna owned a one-bedroom apartment. Her newlywed 20-year-old daughter moved into the apartment with her husband. Anna slept on a sofa in the kitchen. The newlyweds took the bedroom. All went well for the first year until one day, while the daughter was at work, the husband caught Anna coming out of the shower and brutally raped her, filming the act, then told her to "get the fuck out" of the apartment. When Anna tried to contact her daughter, the daughter refused to talk; hung up the phone.

Anna ended up on the street.

"There's nowhere to go," she said. "It's hard on the street... nobody fucking needs you...You've already crossed the line...you're already bad by definition."

Army veteran Sasha's mother was a heavy drinker. She left Sasha with his father, a soldier. After serving in the army, Sasha became a plumber for the emergency services. Then, Sasha's father died. Sasha was 24. He had no family, nobody close, no one to talk to. He started drinking— developed a drinking problem. At time of the interview he had been living on the street for 11 years, spending winters on a blanket laid atop a snowdrift in a strip of forest some three kilometers from where he met Kravtsova for the interview.

Kravtsova cites Igor Antonov, who's worked with homeless people for years:

> It's very difficult to catch someone right at the moment their life is falling apart— And it's extremely difficult to help someone who's already been on the street for a long time. People change dramatically out there — that's what nobody realizes.

Even with incentives to get off the street, the homeless all too often confront prejudice and systemic barriers. In the US, for instance, low wages and high cost of rentals are significant barriers. Millions of low-wage workers, seniors, and families are hard-pressed to find affordable rental homes.[147]

One could say that homeless people are a waste product of economic systems that incentivize money and "success" over compassionate caring. Given those priorities, there's little incentive to provide low-wage workers with a living wage and affordable housing or to invest in mental health and addiction services. A privileged few win. Many struggle to get by. The hardest hit live on the street or die deaths of despair.[148]

Case in point— the second story that caught my eye this morning: "Hotshot" Wall Street lawyers win eight-figure pay packages through bidding wars among top law firms. I'd be tempted to say more power to them but, "many of these new heavy hitters have one thing in common: private equity."[149]

Private equity.

In his book *Plunder: Private Equity's Plan to Pillage America,* Brendan Ballou, a federal prosecutor who served as special counsel for private equity in the Justice Department's Antitrust Division, vividly dissects the predatory business practices of many private equity firms.[150] The plan? Some variation of:

— Borrow money to buy retail and restaurant chains, hospitals, dental practices, nursing homes, and other businesses with lootable assets

— Saddle firms they've purchased with debt and management fees

— Sell and lease back material assets such as real estate to distribute dividends to investors, thus putting hard drain on client cash flow

— Cut maintenance and research budgets to the bone

— Discharge employees

— Loot pension funds, often driving their struggling clients into bankruptcy to do so[151]

...all in the name of "efficiency."

But, as Ballou details, money is the more trenchant incentive:

Management fees: A 2 percent fee on a $1 billion fund returns $20 million to the private equity firm

Transaction fees: Often 1 percent of the deal amount

Performance fees: As much as 20 percent of an investment's revenue

Carried interest: Federal tax rate on carried interest is 20 percent; the top tax bracket is 37 percent

Dividend recapitalization: Special dividends drawn from new debt imposed on the acquired companies and paid to investors who help fund the acquisitions

Over the decade leading up to 2023, private equity firms were responsible for nearly 600,000 job losses in the retail sector and some 20,000 deaths in nursing homes.[152] In 2023 US private equity portfolio company bankruptcies hit a record high.[153]

In short, private equity firms extract millions of dollars in revenue and profit whether the companies they invest in survive or not. And all too many of the companies they invest in fail, resulting in loss of jobs and value to customers and communities.

One can say that relentless, even ruthless, incentives to make money are predominant drivers of our under-regulated capitalist economic system.

Incentives motivate intent or action. They range across the entire spectrum of human needs and wants, from hunger pangs to desire for world domination— e.g., whatever floats your boat and provides reason to get up in the morning.

Incentives are many— basic survival, curiosity, adventure, social recognition, religious and moral beliefs, altruism, family obligation, status, honor, glory, individual or team success, naked greed, coercion as when a robber says, "Your money or your life."

One wonders about incentives to deny scientific evidence of climate change. As I write, the president of the United States has ordered all mention of climate change deleted from federal documents and all federal alternative energy resources frozen.[154] This, despite the fact that...

> In 2024, there were 27 individual weather and climate disasters with at least $1 billion in damages, trailing only the record-setting 28 events analyzed in 2023. These disasters caused at least 568 direct or indirect fatalities, which is the eighth-highest for these billion-dollar disasters over the last 45 years (1980-2024). The cost was approximately $182.7 billion.[155]

If consensus scientific projections aren't enough incentive to curtail greenhouse gas emissions, one would think that ever more frequent extreme weather events would provide a sense of urgency.

This is no doubt true for many. But, for others, short-term fossil-fuel profits and political brownie points tip the scales toward denialism and climate policy obstruction.

Climate physics and chemistry, however, are agnostic to human needs and desires. Unless we curtail greenhouse gases and learn to live within the natural resource doughnut, we are in for ever more climate-related death, material loss, and existential pain. We ask again, can we achieve necessary alignment of incentives within the ticking time bomb of climate change?

Presumably, a primary incentive motivating a majority on Earth is long-term human survival— cultural and economic sustainability. In this idealized world, the seas, forests, and arable lands satisfy essential needs and more for all of mankind. But what can we do about those driven to excessively exploit natural resources by ambition, greed, revenge, or power? How can we incentivize them to live and work within the reproductive capacity of Earth's natural resources?

Regrettably, climate-change deniers seem to be motivated by short-term interests or magical thinking— that is, blind faith that technology or divine intervention will come to our rescue. If their lives are prosperous today, what incentives do they have to believe that physics, chemistry, and the reproductive capacity of natural systems will determine climatic conditions for their children and future generations? Perhaps deeper psychological influences such as fear and perceived helplessness are in play as well.

So we ask, what incentives might motivate climate deniers to embrace science-based public policy and long-term thinking?

Carrots? Sticks?

As nature reminds us with ever more extreme weather events, fire, flood, and drought are blunt motivators. Powerful public push back against climate-change denial would be far more timely and humane. We must demand that influential deniers and their political and media enablers get behind constructive climate policy or get out of the way.

Heady challenge. What to do?

What can we do?

Makes my head ache.

9

Work

Zip back in time. Step into the shoes of an 18th-century Russian serf. Your work follows seasonal rhythms. You and your family must plant, cultivate, and harvest rye, oats, and a few garden vegetables. Care for cows, pigs, and horses. Backbreaking, relentless labor. You must produce enough to meet your family's basic needs and put away for frigid winter, despite vagaries of weather, war, and corrupt officialdom. And more, substantial fruits of your family's labor must go to the feudal landlord and tax collector. You labor in debt bondage, resistance brutally suppressed.

Teleport to England.

The Enclosure Acts passed by Parliament have allowed wealthy landowners living in London to fence in open fields, fens, marshes, rocky land, and moors that your family has accessed for generations to pasture animals, harvest meadow grass, fish, and collect firewood. Your family is shivering in the cold; facing starvation. Work is available in factories for you, your wife, or your children. It's harsh, dangerous work, 12 to 16 hours per day, six or seven days a week, for meager wages. But you have no choice— you must move off the land.[156]

Life is little better for the weavers and skilled craftsmen in your area. Factory owners want cheap labor and people are easily replaceable.[157]

Now imagine your life as a slave on a pre-Civil War Florida panhandle plantation. The owner needs cheap labor to profit from his land. The incentives he inflicts to keep you working in the hot Florida sun are brutal.

Slavery has existed on nearly every continent and throughout most of recorded history. Though outlawed around the world today, slavery persists with an estimated 50 million people trapped in slavery worldwide.[158]

For many in the workforce today, freely engaged paid employment is no escape from oppressive work conditions. Summer 2023 was the hottest in 2,000 years.[159] Nevertheless, despite an 88 percent increase in heat-related deaths in Florida from 2019 through 2022,[160] Florida Governor Ron DeSantis signed into law on April 11, 2024 a measure intended to prohibit local workplace heat standards that go beyond state and federal laws. The law bans cities and towns from:

— Regulating drinking water, cooling measures, or recovery periods

— Posting or distributing materials informing workers how to protect themselves

— Requiring first aid or emergency responses[161]

Given Florida heat-death statistics, it's evident that the state and federal laws were inadequate to protect workers.

Did Governor DeSantis not step outside into 95-degree noon-day sun nor check with his Department of Public Health? Or— did he just not care?[162]

The Associated Press reported on May 31, 2024 that:

> ...death certificates of more than 2,300 people who died in the US (in summer of 2022) mention the effects of excess heat, the highest number in 45 years of records.[163]

Provisioning the basic needs and near infinite wants of humanity takes work, labor. Many people find creative or self-fulfilling satisfaction in work. For them, work is its own reward. Some dedicate their lives to working with the poor, young children, frail elderly, or the environment out of faith, compassion, or desire to contribute to the greater good. More than 100 million open source software developers give away the fruits of their labor on GitHub at no cost.[164] Their work powers the Internet. Fine artists and novelists work long hours in obscurity without pay. But for many, work is tedious or exhausting and simply a means to afford basic necessities. Some are trapped by debt in jobs they'd rather not do.

Projects of scale require coordination and administration of a workforce. CEOs of transnational corporations get big bucks for their political and management skills:

> The American worker's average annual wage in 2023 was only $65,470 USD, according to the Bureau of Labor Statistics. At

that wage, it would take 445 years to earn as much as the middle-of-the-road chief executive.[165]

CEO incentives in many corporations seem to prioritize profit, stock appreciation, and top management pay over average employee contribution to corporate success. Recent events involving a once premier aircraft manufacturer[166] and, in another case, a leading processed meat provider, suggest that management prioritized profit over human lives.[167]

Many jobs are physically taxing, tedious, distasteful, or dangerous. Who can be found to do such jobs? What's the incentive? Sufficient pay to make ends meet is often enough. High pay may induce many to take on dangerous jobs, but many high-risk jobs are not all that well paid. Some low-skilled workers take whatever job they can get. But, what about employers? Why are many motivated to pay as little as possible for labor?

You know the answer. It's taught in Econ 101:

$$P = I - E$$

Profit equals Income minus Expenses.

Labor is an expense. So cheap labor boosts profit. Throughout history, oppressive measures, violence even, enforced by law, have been go-to practices in pursuit of profit. The profit may not even be monetized. In pursuit of adulation and glory, pharaohs and monarchs enforced slave labor with oppressive measures to construct pyramids and provide temple services such as planting and harvesting crops, tending cattle, and weaving textiles.

So one can't talk intelligently about work without thinking hard about power—the balance of power between employer and employee, that is, working conditions and allocation of profit.

Consider another "law" taught in Econ 101: Supply and Demand.

In thriving economies labor demand pushes up wages. But during recessions and depressions employers lay off workers, pare down wages. Families scrimp to make ends meet, perhaps even go hungry.

Or worse.

Economist Anne Case and Nobel Prize winner Angus Deaton have analyzed and documented the fall of life expectancy in the US in recent years. They attributed reduced life expectancy to growing power of corporations, the weakened position of labor, and rising health-care costs. While capitalism has lifted countless people out of poverty, they say, more recently it has blighted the lives of blue-color workers.[168]

Consider the human consequences underlying the abstractions of Econ 101.

Oxfam tells us that:

- Living standards around the world were more similar in the early 1800s than they are today.
- Global inequality grew substantially after the Industrial Revolution.[169]
- Today, the 10 richest men in the world own more than the bottom 3.1 billion people.[170]

Would we choose to return to economic realities of the early 1800s?

I wouldn't.

But work of the world must go on. Even the poorest of the poor must work one way or another to secure basic needs.

Dense urban societies today require interdependent coordination of physical labor, innovation, finance, trade, transport, governance, and administration. Contemporary society would be seriously impoverished without education, security, conflict adjudication, law enforcement, sanitation services, and health care. The list of tasks that must be competently executed every minute of every day to keep contemporary society humming is beyond comprehension. The know-how and talents required are equally vast.

There's little reason to believe that this will change in a doughnut world. But jobs in extractive industries and manufacturing will most certainly change, with possible loss of mining, logging, and wild-catch fishing jobs. Consumption of material goods will likely decline with impact on trade, transport, and advertising. Emerging technologies will likely eliminate yet more jobs in diverse sectors.

But jobs lost in one sector may well shift to others. Job opportunities in resource monitoring, management, and enforcement may replace jobs lost in extractive industries. The International Energy Agency (IEA) tells us that 32 million people worked in the fossil fuel industry in 2023.[171] Most of those jobs would not be on offer in doughnut economies. However, clean energy sectors employed 36.2 million workers in 2023.[172]

Clean energy jobs will certainly become ever more available, as will work on projects directed toward recovery of ecosystems that have been degraded, damaged, or destroyed.

Resource substitution will open additional opportunities. Engineered mycelia composite, for instance, can replace many construction materials, including door cores, panels, flooring, and foam, wood fiber, and plastic insulation.[173]

Manufacturers will be called upon to produce more durable, repairable, and recyclable products with consequent reshuffling of job responsibilities and titles.

One can imagine demand for repair and restoration services flourishing— everything from consumer electronics to electric cars.

Moreover, human curiosity and imagination are without limit. So jobs in research, design, healthcare, education, and entertainment sectors will likely thrive. Banking, finance, law enforcement, and public service will certainly continue to provide jobs for many.

The issue of equitable and just allocation and administration of jobs and wages, however, is very much on the table. Run your finger down a GDP-by-country list and economic disparities are glaring.[174] Disparities are equally glaring within the US— compensation ranges from less than $ 7.25 USD per hour for hourly workers to nine-figure pay packages for celebrities, athletes, and corporate CEOs.

No doubt supply and demand is a significant driver of wage disparities Arguably, athlete LeBron James, CEO Larry Ellison, and singer Taylor Swift are one of a kind. But why are child care workers and preschool teachers paid less than 97 percent of all other occupations in the US?[175]

Why do millions of low-paid workers in the US rely on the Supplemental Nutrition Assistance Program (SNAP) to put food on the table,[176] while "hotshot" lawyers earn top-dollar to help private equity firms plunder the US economy?[177]

The question of equitable and just allocation and administration of jobs brings us back to metrics and incentives

1. What are the most appropriate metrics for allocation of jobs and wages in a doughnut world?
2. What are the most appropriate incentives?

And this, in turn, leads us back to values and beliefs— culture Eighteenth century Russian nobles and US Southern plantation owners profited from ownership of serfs and slaves with hearty application of caning and the whip. Marxists would argue, "From each according to his ability, to each according to his needs." In capitalist cultures, many believe that supply and demand is the way to go.

It's easy to take for granted the millions of low-wage workers who contribute to the well-being and comforts of our lives— sanitation workers, hospital orderlies, preschool teachers, restaurant waitstaff— the list goes on. For many, supply in excess of demand dictates less than living wage. We conveniently forget the comfort, convenience, and security that essential workers contribute to our lives.

Imagine a world where all low-wage essential workers worldwide take year-long vacations— all at once. How would this impact quality of life in New York City, Paris, or Kyoto?

Now imagine a world where all eight-figure "hot-shot" lawyers working for private equity firms are transported to a distant galaxy by aliens.

Which world would you rather live in?

So why do we accept orders-of-magnitude discrepancies in pay? Oh, yes, forgot— the number of low-wage laborers available to fill limited jobs, that is, supply and demand.

As it stands, many wage earners work long hours or multiple jobs to make ends meet. They have little time for family, cultural, or civic engagement. Menial but essential work is often disparaged and underpaid. Income and wealth disparities within and across national boundaries are staggering, as are wage earner incomes across commodity export and commodity import nations.

Then there's sense of contribution. In his book *Bullshit Jobs: A Theory*, David Graeber tells us that many jobs in our developed economies are meaningless, unfulfilling, and perceived as useless by a significant segment of the workforce.[178]

Questions:

- Is this simply the natural order of things?
- Or, is it the result of exploitative power relations within and across our economic systems?

Opting into a doughnut economy is an opportunity to reward work for value contributed rather than vagaries of supply and demand.

Aspiration: Imagine if we could assure full employment; sustainable livelihoods; more time for family, leisure, and civic activities; and high quality of life for all.

But how?

Many ideas are already on the table:

— Guaranteed basic income[179]

— Living wage pay scales[180]

— Four-day work weeks[181]

— Employee ownership[182]

— Worker cooperatives[183]

— Extended parental leave[184]

— Elimination of child labor[185]

— Community-based natural resource management[186]

— Labor representation on corporate boards[187]

— Break up of monopolies and oligopolies[188]

— More effective regulation of financial speculation[189]

— Elimination of inequalities embedded in tax policies[190]

— Increased inheritance taxes[191]

— Elimination of sweatshop and slave labor[192]

No doubt there are many more.

How realistic?

- Mondragon Cooperative (MC) is considered one of the most successful examples of worker cooperatives in history.[193]

- Living wages result in more motivated and productive workforce and lower staff turnover.[194]

- Japan is urging citizens to try a four-day work week.[195]

The list goes on.

But, still, we must ask:

- Can doughnut economies provide sufficient food and materials to satisfy basic life-sustaining needs for all?

- Can we equitably allocate and distribute natural resources without compromising nature's resiliency?

- Can we eliminate exploitative employer/employee relationships such as slavery, sweatshop work conditions, and less-than-living-wage jobs?

- Can we encourage and support every child to rise to fullest potential and contribute personal best for the greater good?

- Can we make work more rewarding and meaningful for all?

It all comes down to will, incentives, culture, technology, and governance.

10

Technology

Climate change?

Biodiversity loss?

Toxic waste?

Water scarcity?

Resource depletion?

Deforestation?

Some say that artificial intelligence (AI) will save the day. Color me skeptical.[196] What makes us think artificial intelligence will be any wiser than the AI designers and programmers who created it or the data scraped off of the Internet and other sources that trained it?[197]

One can argue that five serendipitous drivers fueled our journey from the savannas of Africa, mountains of Europe, and banks of the Yellow River to our tech-empowered mega-cities of 2025 CE:

- Curiosity
- Imagination
- Tools
- Know-how
- Generational transmission of knowledge

Curiosity motivates investigation of "Why?" and "How?" Imagination envisions "What if?" Tools extend our hands and senses; enable us to reshape our environment. Know-how enables us to use tools efficiently and effectively. Generational transmission enables us to pass hard-won insights and skills onto our children and others.

Technologies are integrated systems of tools and know-how. They inspire creators to craft innovative works of art, manufacturers to transform raw materials into valuable products, and scientists to see ever deeper into atoms, ever more insightfully into human physiology, and ever further into outer space. Each advance spurs new vistas and opportunities.

We started with little more than chipped stone and broken branches; now deploy interstellar spacecraft, communication satellites, server farms, high-speed rail, precision machine tools, massive earth movers, robots, household appliances, and more.

But technology is two-faced. Tech can be an agent of creation or destruction; help save lives or wreak near instantaneous mass slaughter.

Consider fossil fuel technologies— on the plus side, they have provided unprecedented comfort, convenience, and mobility. But we know now that waste products of fossil fuel combustion — CO_2, NO_2, and others — are threatening marine and terrestrial life on Earth, including our families, friends, and the rest of humanity.

Or consider nuclear technologies— early proponents promised energy too cheap to monitor. But among the first applications were bombs dropped on Hiroshima and Nagasaki, which killed some 100 to 200 thousand civilians and provoked a nuclear arms race.[198]

We've lived under threat of nuclear holocaust ever since.[199]

The peaceful applications of nuclear technology have a troubled history of accidents— Three Mile Island, Chernobyl, and Fukushima among others.[200] Disposal of radioactive nuclear waste is yet another on-going concern.

Absence of caution in the rush to profit from the Internet provides another example. Security analyst Davi Ottenheimer lists tragic outcomes:[201]

— Lynchings coordinated on WhatsApp in India[202]

— Ethnic violence enabled through Facebook in Ethiopia[203]

— Mass shooters inspired and motivated through YouTube's radicalization pipeline[204]

— Self-harm content pushed to vulnerable teens on TikTok[205]

— Twitter (now X) pushing Nazi propaganda[206] and allegedly enabling coordinated harassment campaigns to encourage suicide of LGBTQ individuals[207]

Sextortion of minors via social media and email is a growing concern. The US Attorney's Office tells us that:

> Over the past year (2023), law enforcement agencies have received over 7,000 reports related to the online sextortion of minors, resulting in at least 3,000 victims, primarily boys. More than a dozen sextortion victims were reported to have died by suicide.[208]

Even now, we're seeing history repeated in the rush to roll out AI. According to a report by Goldman Sachs, "...mega tech firms, corporations, and utilities are set to spend around $1 trillion on capital expenditures in the coming years to support AI (artificial intelligence)."[209]

But...

— Tesla AI algorithms are causing hundreds of sudden "veered" crashes, killing dozens of passengers.[210]

— The ability to create new content such as conversations, stories, images, videos, and music through generative AI contributes to divisive disinformation and polarized public debate. Deepfakes (AI content that makes it appear as if someone is doing or saying something they have not actually done or said) are exploited to convey disinformation, influence public opinion, and sway elections.[211]

— A Gallup survey finds that more than a fifth of workers in the US worry they might lose their jobs to generative AI.[212]

— Goldman Sachs estimates that the data centers that host AI will demand 165 percent more power by 2030.[213] The technology editor of Yahoo Finance, Daniel Howley, tells us that "Big Tech is going all in on nuclear" to meet AI power requirements.[214]

The World Nuclear Association reports that 440 nuclear power reactors are currently operating in 31 countries, some 70 are under construction, and 100 more are planned.[215] What could go wrong? Three Mile Island, Chernobyl, and Fukushima are more than ample reasons to give us pause.[200]

Optimists contend that AI and nuclear energy are boons to mankind. Others express serious concern over unintended environmental consequences. There's money to be made in AI, so uncritical boosterism is rampant. A useful distinction can be made between machine-learning models trained with verifiable data and applied to well-defined tasks (Narrow AI) and General AI trained with data indiscriminately scraped from the web and promoted as capable of human-like intelligence.[216]

For a critique of General AI, check out Stephen Fry's essay:[217]

AI: A Means to an End or a Means to Our End?

https://stephenfry.substack.com/p/ai-a-means-to-an-end-or-a-means-to

Post Carbon Institute Senior Fellow Richard Heinberg's AI concerns are also worth thought and debate:

Why Artificial Intelligence Must Be Stopped Now

https://www.localfutures.org/why-artificial-intelligence-must-be-stopped-now/

This is not to say that we should cease development of AI or new technologies across the board. Rather, the question is, how can we weigh trade offs, foresee serious harms, and minimize unintended injury and loss due to misapplication?

We might start by considering how emerging technologies convey power— to whom? To do what? Alternative energy technologies, for instance, threaten the global fossil fuel industry. Since alternative energies are essential to the long-term sustainability of life on Earth, it's prudent to think hard about how to manage the transition from fossil fuels to alternatives and how to wisely craft incentives and support to compensate for loss of investment and livelihoods. We should certainly consider the long-term implications for human sustainability. And, in all prudence, we should thoughtfully develop global standards and regulations.

In short, the environmental and social implications of emerging technologies at global scale deserve serious thought and open debate. Do they harm or support the natural environment? Do they contribute to the well-being and social cohesiveness of all on Earth? As matter of policy, should emerging technologies be kept secret for reasons of national security or corporate gain, protected by patents and copyright to be bought and sold as intellectual property, or freely and widely disseminated for benefit of all?

Looking forward, if our intent is to provide sustainable abundance and well-being for all while respecting and enforcing natural constraints, what technologies will serve us best?

Foremost, our economies must provide:

- Clean water
- Nutritious food
- Secure housing
- Comfortable apparel
- Convenient transport
- Accessible medical care

They must maintain and improve:

- Sanitation systems
- Physical infrastructure
- Educational, healthcare, and scientific facilities
- Civic and commercial buildings
- Factories and distribution systems
- Libraries, museums, theaters, and stadiums
- Parks, reserves, and recreational facilities

Questions:

- How can we best mobilize and advance our tools and technologies to improve and sustain the quality of life for all while preserving the regenerative health and capacity of our natural ecosystems?
- How can we deploy technologies to help workers rise to fullest self-potential, to find significance and meaning in what they do, in what they accomplish?
- How can we regulate AI in support of sustainability goals and social justice— curtail spread of disinformation, protect personal privacy, and prohibit unauthorized use of literary and creative works as AI training data?

Think of technology as a vast palette of creative options for designing, implementing, and maintaining a sustainable future for all life on Earth.

Possibilities and opportunities come down to imagination and incentives. What we can imagine, we can do for the good of mankind— or not. Money, wealth, and power have motivated our drive to dominate nature and to enslave and exploit fellow human beings. The relentless drive for money, wealth, and power is navigating us toward a disastrous dead end. But imagine healthy natural systems worldwide, abundance for all, and a sustainable future. What to do?

What can we do?

Many technologies are under development or already deployed to reduce greenhouse gas emissions and revitalize degraded ecosystems:

- Satellites and remote sensing devices monitor environmental conditions on Earth and track extreme weather events.
- Grid-scaled solar, wind, and battery power plants,[218] electric vehicles[219], and high-speed rail[220] are among the most obvious technologies to reduce greenhouse gas emissions but, in the US, confront significant pushback from the fossil fuel industry.
- Global agricultural food systems contribute an estimated 34 percent of human-caused emissions of carbon dioxide, methane, and nitrous oxide. A study by researcher P. Thamarai and colleagues[221] suggests that technologies old and new will be required to achieve long-term agricultural sustainability— organic farming,[222] precision farming,[223] vertical farming,[224] and carbon sequestering,[225] among others.
- Cement is responsible for 8 percent of global carbon emissions. Limestone calcined clay cement, algae-grown limestone, electric recycled cement, and other technologies promise significant reduction of carbon emissions.[226] The international nonprofit organization Concrete Group's ConcreteZero initiative has set a goal of net zero concrete emissions by 2050.[227]

In short, technologies help us live in harmony with nature or alienate us from both nature and each other. Every advanced technology is an ecosystem cultivated by vision, intent, scientific research, investment of time and money, and application.

Companies often develop and market products with insufficient understanding of long-term environmental fallout. Think carcinogenic pesticides and widespread use of "forever chemicals." Some companies even know of harmful effects but keep the facts secret.[228] Many technologies of war are advanced in secret with little thought to misapplication or concern over unintended proliferation, global stability, or environmental impact.

In a doughnut world it's crucial to understand intent and potential harms of new technologies.

Yet more questions:

1. How does the technology advance our understanding and stewardship of the natural world?
2. To what extent will the technology degrade or deplete natural resources?
3. How should we monitor application of the technology?
4. How would we mitigate or reverse unintended consequences?
5. Who would be held accountable for unintended consequences?

Food for thought:

— Throughout much of the 20th century, countries have invested heavily in technologies of war. Our inability to rein in unaccountable power through negotiation and compromise has cost millions of lives.

What have we learned?

How can these lessons help us envision and manage a sustainable future?

— Chinese scientists recently reported creation of a robot controlled by a blob of human brain cells.[229]

What are the promises and perils of this research?

— The Rocky Mountain Institute claims that carbon capture, biomass, and blue hydrogen technologies have "repeatedly under performed," obscuring the success of solar, wind, batteries, and other alternative energy solutions.[230]

What do you think?

11

Media

Our journey toward sustained global habitability requires consensus and cooperation at global scale. We need to share ideas and technologies without limit. We need to base the many decisions and choices we face on sound, replicated science. This requires universal understanding of that science and more. It requires tolerance, mutual respect, and deep understanding of human nature and incentives to bring out the best in us.

Fractious Internet communities stand in our way— bubbles of ignorance, self-dealing, and malice.

Perhaps you've noticed. There's a worldwide war raging across our media of mass communication. Prime targets— your mind and mine. Entities seeking control over our attention, time, beliefs, values, opinions, money, and behavior include nation-states, transnational corporations, cybercriminals, electoral candidates, TikTok influencers, local pizza shops, and more. Some aggressive media sources assault us head on with blatant lies and misinformation. Others are more duplicitous. Some work to serve our best interests. Others most definitely do not.

We ask much of our media of mass communication— newspapers, radio, TV, social media:

- Information
- Early warning of impending threat
- Entertainment
- Education

- Connection

We depend on media to serve as:

- Public forum
- Watchdog over government, business, and civic institutions

The mass media extend our innate curiosity to realms beyond our immediate surroundings— community, nation, global affairs. Indeed, we need to know what's going on around us. In some circumstances situational awareness is a matter of survival. It pays to step carefully in rattlesnake country.

When we're bored, we seek stimulation. The mass media comply with comedy, vicarious adventure, and drama.

When we're beset or bereaved, mass media provide useful perspectives, facts, and recommendations; remind us that we're not alone.

Responsible reporting and media scrutiny help hold politicians and governments to account— an essential role in vibrant democracies.

Within recent memory mass media was structured around a limited number of one-to-many channels, some continent spanning and others local. Communication technology was expensive. Ownership of major radio, newspaper, and television outlets was in the hands of wealthy individuals and corporations— and still are. Editors and presenters were seen as trustworthy, authoritative figures. Newspapers of record and nightly TV news provided information and perspective that tended to unite us. It can't be denied, however, that traditional mass media promoted the views of power elites.

Access to traditional media content was either inexpensive or free, subsidized by advertising. But conflict of interest was inherent. Advertising is a benefit to the extent that it informs us of useful products and services. But advertising is parasitic. It craves dominance over attention.

Quality newspapers enforced editorial walls between news and advertising. Broadcasters were licensed by the US Communications Act of 1934 to reduce interference across the broadcast spectrum. The act required broadcasters to operate in the "public interest, convenience, and necessity."[231] The Fairness Doctrine was adopted by the US Federal Communications Commission in 1949 to ensure that broadcasters presented "...controversial issues of public importance in an honest, equitable, and balanced manner."[232] News was seen as public service by early broadcasters. But circulation and ratings dictated whether for-profit media companies thrived or died. The Fairness Doctrine was repealed in 1987, during the Reagan administration.[233] News became infotainment;[234] entertainment, a vehicle for product placement.[235]

The Internet "democratized" mass media by proliferating cheap many-to-many channels of communication. Anyone with a modem or router and a personal computer could become a content creator, editor, or presenter. The Internet was seen at inception as bringing enlightenment and institutional transparency to all.

We know now how the hopes of Internet visionaries have turned out. The Internet is the global nervous system of commerce. But agency of enlightenment and transparency? Not so much.

The Internet enables work from home, on-line shopping, email, video chat, telemedicine, and other conveniences. More than five billion people are connected.[236]

But there are dark demons lurking just beneath clacking keys, poised to pounce from behind our touch screens.

Tech innovators created social media silos. Their business models provide easy access to the Internet in exchange for personal information, which tech giants suck up and monetize with every keystroke.

Personal data files are bought, sold, traded, hacked, and transferred across the Internet. Troves of files containing telephone numbers, birth dates, social security numbers, recent purchases, time-at-location data, and Internet search histories abound in corporate and hacker databases.

Cybercriminals gain access legally and illegally, hijack our personal information, and use it to defraud us. They break into defense, finance, healthcare, and household computers to hold vital information for ransom and to further profitable scams. Globally, cybercrime is projected to cost $12 trillion in 2025.[237]

Social media has also become a powerful tool for mass surveillance, social control, and repression. According to a 2019 survey released by Freedom House:

> 47 of the 65 countries assessed featured arrests of users for political, social, or religious speech— a record high. The blanket monitoring of online activities for undesirable or illegal speech will undoubtedly lead to more arrests, particularly in environments that lack strong protections for free expression.[238]

It's been argued that personal privacy is dead. It's indisputable that disinformation, propaganda, and suppression of speech flourish.

In democracies, everyone is entitled to their own opinions. But social media "influencers" broadcast personal opinions and, often, misinformation and unfounded malicious rumors to tens, thousands, even millions of followers. State actors and ideologues filter, amplify, and weaponize topics that resonate with their worldview or propaganda objectives by clicking on "Like" buttons and reposting links to their contacts and media outlets.

State and non-state troll farms flood social media with misinformation, disinformation, and outright lies with intent to incite civic division and chaos among targeted opponents.[239]

Social media companies monitor clicks and likes and feed results into algorithms that further amplify selected topics to promote "engagement" and advertising revenue. Media reporters and producers pick up Internet "buzz" and report it to their readers, listeners, and viewers.

Clickbait rules.

Official and unofficial fact-checkers embedded throughout the media ecosystem strive to correct information that they deem as incorrect. Ideologues demonize the fact-checkers and ridicule opinions and ideas they disagree with— perhaps even threaten violence. Some incite literal violence. In early 2025, Meta announced that it was ending its fact-checking program on Facebook, Instagram, and WhatsApp. Rather, it would allow users to correct inaccurate and false posts.[240]

At the same time, autocratic countries censor dissent and block access to the global Internet with powerful network-monitoring tools and firewall technology.

Political discourse has become toxic. The juvenile insult-and-trashposting duels of video gamers have been weaponized and widely deployed for hire as malicious online harassment, threats of violence, and political smear campaigns. Death threats against judges, legislators, and election workers in the US have become all too common.[241]

Some such attacks are launched to induce fear, self-censorship, and political apathy to the benefit of ideologues and corrupt regimes. Internet search engines early on served us with information relevant to our needs. Now they serve us up to advertisers— distracting us with advertising that wastes our time. Indeed, the FBI recently warned that cybercriminals are seeking to defraud search engine users by impersonating commercial brands.[242]

The global information ecosystem today is fragmented, polarized, and noisy. Trust in science, credible expertise, and governing institutions is diminished.[243] Consensual truth has given way to "truthiness."[244]

Anne Applebaum sums it up well:[245]

> The Internet...fundamentally changed the way that people understand the world, particularly the political world and political information. So the way that people now get information is through short bursts of messages on their phone. There has been a kind of dumbing down of everybody.[245]

In short, the dream of universal enlightenment is far from reality.

Cleaning up our polluted Internet discourse is a challenge of the same order and urgency as elimination of greenhouse gas emissions and toxic waste.

But how can it be done?

The Union of Concerned Scientists says, "We all have the power to counter and interrupt disinformation."[246] Finger pointing doesn't work, they argue. The way we consume and respond to information on the Internet, however, does:

- Do not share disinformation.
- Alert others to watch for and expect disinformation on a specific topic.
- Amplify accurate stories and information from trusted and credible sources.
- Report disinformation.
- Build relationships through one-on-one conversation.

I would add.

– Consider the source of extravagant claims. What are the experience, credentials, and reputation of the source? How does the source benefit by making such claims? Incentives for social media influencers to post clickbait and propagate extravagant claims include increased engagement, clout, "Likes," and income from sponsorships, subscriptions, and product sales.

— Consider who benefits by persuading you and others that an assertion of fact is true. Distrust of government, cultural institutions, science, and qualified experts benefits autocratic wannabes and adversarial geopolitical actors, as do civic polarization, confusion, and political apathy.

— Seek facts from multiple credible sources and contribute to Internet discourse with integrity, truth as your standard, and respect for other points of view.

The UN has published *The United Nations Global Principles for Information Integrity* to help make Internet discourse more trustworthy. It warns that misinformation, disinformation, and hate speech threaten democracy and human rights and undermine public health and climate action.

UN principles include:[247]

- Societal trust and resilience
- Healthy incentives
- Public empowerment
- Independent, free, and pluralistic media
- Transparency and research

Consider:

Standing atop a skyscraper you can believe anything you choose. But jump off and gravity rules. Rumor mongering and deceit may be embedded in human nature. They may sustain profitable Internet gigs. But climate change and ecosystem destruction are facts of nature.

We ignore them at our peril.

Our communication technologies are powerful tools for conveying existential facts of nature and what we must do to thrive within the limits of nature's regenerative capacity. It's up to us to use both traditional media and the Internet in the best interests of our natural environment— in the best interests of humanity's future on Earth.

We have much to gain. A habitable Earth to lose.

12

Governance

Ah, here we confront mist-shrouded forking paths. It will take, I fear, legions of political scientists, principled politicians, diplomats, and all hands on deck to dispel the toxic fog of distrust, intolerance, fear, greed, and power rivalries that endanger life on our planet.

Here be dragons!

Consider:

— Citizens worldwide are sorted into 195 kleptocracies, autocracies, theocracies, monarchies, authoritarian dictatorships, compromised and thriving democracies. Each nation-state is distinguished by history, culture, and political and religious factions.

My country, most economically prosperous of the lot, is socially and culturally polarized and politically gridlocked.

— Wars rage in several regions of the world; a few countries suffer bitter civil war. Yet somehow we must join hands around the globe to work together toward the goal of sustainable ecosystems and abundance for all.

— Legal prohibition of desirable goods inspires criminal traffic.

A 2025 report from Washington DC think tank Global Financial Integrity estimates that illegal fishing, logging, mining, and wildlife trade generate collectively from $155 billion USD to as much as $383.4 billion USD annually.[248]

— In the decade between 2012 and 2023, more than 2,100 environmentalists were killed trying to protect their land and resources.[249]

— People rebel for reasons petty and large. Remember resistance to COVID masks? And oh, yes, let's not forget the January 2021 storming of the US Capitol nor recurring bloody coups, rebellions, and revolutions that stretch back through history.[250] We must listen with open heart to one another, negotiate differences.

— Nor should we forget that brutal, charismatic autocrats and dictators, mass-murderers of millions, crop up much too often throughout history.[251] Bloody campaigns of imperial conquest, rape, slaughter, and pillage are a recurring theme.

Yes, *Homo sapiens* can be a nasty lot.

If we can't govern our own emotions, impulses, and appetites, then we need systems of social policy formulation and implementation, laws and law enforcement, and defense from aggression of geopolitical rivals— government.

Government is the consolidation of power and the means to deploy it.

Questions:

- Who holds power to do what and how are they held accountable?
- What theories, structures, and scale of government are best suited to manage sustainable natural resources and a good life for all?

We face obstacles and challenges, yes. Big time. But in daily life we see love, compassion, generosity, sacrifice, and courage. We see people living and working together in harmony. Harmonious human relations are the norm and mass violence the exception. This is bedrock upon which we can build.

For us to continue on our heedless way and let nature, ever more exploited and polluted, forge a path for us is not an option. We must choose a fork toward sustainable economics and lifestyles. What to do?

What must we do?

Somehow we must step away from autocratic, kleptocratic — even jingoistic — control and distribution of resources. We must move on global scale toward policies and alliances that work within nature's rhythms and interdependencies to support just and equitable quality of life for all.

In a narrow sense, sustainability and doughnut economics imply zero-sum natural resource allocation. If Timberman Tom cuts 100 trees, that many fewer remain for Woody Woodsman. This could well be a recipe for conflict and violence. Worst case, a Mad Max dog-eat-dog struggle for dwindling natural resources. But that's not an acceptable option. Neither is overwhelming nature's regenerative capacities. In either case, there's a bleak future for humanity. There must be a better way.

So the question is, how can we move effectively from the idealism implicit in Kate Raworth's doughnut model to sustainability? What institutions and structures of governance will be required?

Jonathan S. Blake and Nils Gilman published thought-provoking insights in an essay called "Governing for the Planet:"[252]

> Managing world-scale, or planetary, problems requires acting for 'world interests'. Thus planetary problems require solutions at the planetary scale. The scale of these problems is incommensurate with our current institutional capacity to govern them.

But a single world state isn't the answer, they say:

> Rather, we must create institutional structures that foster flexibility, with multiple institutions for multiple scales, individually and collectively crafting effective governance for diverse populations seeking to thrive on one interconnected planet.

Blake and Gilman prefer the word "habitability" to "sustainability" to include care for all living things on Earth.

Just as I write, the G20, a group of the world's largest economies, has launched a Task Force for a Global Alliance against Hunger and Poverty. Members attending the July 24, 2024 ministerial meeting endorsed the Alliance's institutional structure, approving its four constitutive documents.[253]

The goal is to "...better align domestic and international support to help countries build and implement programmes and policy instruments that reduce hunger and poverty."[254]

> Among the ways to address the new ideological, military and geopolitical challenges is to show that multilateralism, even at its most minimal, can work...every country now needs multilateralism.[255]

Human history has spawned countless economic and political systems. Our kind fought devastating wars throughout the 20th century under various ideological banners. We can look to the past, but few systems supporting large populations have successfully balanced sustainability of the natural environment with abundance for all. But some systems have done better than others and utopian ideas abound. Check out William Morris's 1890's novel *News from Nowhere* for a delightful fictional take.[256]

Fact is, we do know how to govern well. Richard E. Blanton and colleagues defined good governance succinctly in their study of why polities collapse:

> Good governance includes the degree to which leadership will accept limits on its power, is willing to develop the governing capacity to identify and punish official corruption, is willing to provide citizens access to an impartial judiciary, and is prepared to implement equitable taxation, to open up access to positions of governing authority without favoritism, and to provide public goods beneficial to all households.[257]

The hard nitty-gritty political work starts with the human heart; spans households, neighborhoods, communities, states, provinces, nation-states, transnational alliances, and NGOs. It involves raising living standards of the undernourished and insecurely housed around the globe, institutionalizing respect for workers who undertake the menial tasks of society, assuring a living income for all, supporting the infirm of all ages and the destitute elderly, addressing the core drivers of domestic violence and drug addiction, reducing material consumption, and restoring blighted natural systems to sustainable regenerative health.

We must work in harmony with nature rather than dominate.

The good news is that:

— We have capacity to learn.

— We're abundantly innovative.

— We expect fairness, even as babies.[258]

— In a crunch, we know how to work together toward the greater good.

What better good than sustainable abundance for all?

For the sake of our children and all who follow, we must demand nothing less.

13

Habitability

> Climate change is an existential threat to humanity. To secure a livable future, each of us needs to ask ourselves: how can I best contribute at this crucial moment in human history?[259]
>
> *Fabian Dablander, University of Amsterdam*

Indulge me, no— think along with me in a speculative venture that will no doubt trigger scorn across the economics profession. I mean, what do we know?

But consider: economists are concerned with distribution of goods and resources in society. We're the subjects of their theories, research, and models. Their work product impacts our lives. Why shouldn't we have input? Besides, trust me, we'll be skateboarding in the slipstream of fully credentialed economists such as Kate Raworth,[260] Herman Daly,[261] Sam Alexander,[262] Lorenzo Fioramonti,[263] Beth Stratford,[264] Simon Kuznets,[265] and others.[266]

Inspiration for our thought-excursion comes from Kate Raworth's doughnut model, which emphasizes the urgent need for sensitive, near-real-time metrics to measure just where we stand with respect to habitability of life on Earth.

So, premise: We need a vivid readout of critical ecosystems and human welfare metrics— a readout as widely reported and policy-relevant as GDP, but as provocative as the Bulletin of the Atomic Scientists' Doomsday Clock. One glance at the Doomsday Clock and you're emotionally invested in geopolitical state-of-play.[267]

Granted, thinking about metrics to assess human welfare and the health of ecosystems is a nerdy endeavor. But, future of life on Earth? It's worth a thought or three, don't you think?

As Lord Kelvin said in the 1880s:

> ...when you can measure what you are speaking about, and express it in numbers, you know something about it; but when you cannot measure it, when you cannot express it in numbers, your knowledge is of a meager and unsatisfactory kind.[268]

Our challenge, then, is to devise an indicator that conveys at a glance the relative abundance of natural resources worldwide, as well as how well the products and services of nature are distributed to satisfy basic human needs.

Questions:

— What should we call our fanciful new indicator?

— How can we visualize it?

Need your input here— but perhaps we might call it Gross Global Habitability (GGH).

GGH metrics must capture the Earth's vast, interrelated, dynamic systems, both natural and human. Let's call the state of ecosystems and biodiversity Gross Global Resources (GGR). And let's call economics and human welfare Gross Human Welfare (GHW).

In the spirit of Don Quixote, we might define GGH as:

$GGH = GGR + GHW$

Any reason why every nation-state shouldn't have it's own Gross National Habitability index? That would certainly highlight areas of the globe most in need of help, as well as provide the data for our global index.

We'll no doubt need experts from many disciplines to do the heavy lifting. But here, perhaps, we can get an overview of work in progress and broad-brush our scope of concerns. Fortunately, economist Herman Daly and others have staked out the foundation.[269]

A working definition of "natural resources" is a good place to start:

> Natural resources are commodities or the raw inputs used to manufacture everything from intermediate goods to finished products.[270]

I would add that natural resources include air, fresh water, phytoplankton, and fertile soil. We might even add nature's contribution to human mental health[271] and its spiritual significance.[272]

Phytoplankton?

Phytoplankton support entire marine and freshwater food webs. They photosynthesize carbon dioxide to produce as much as half of the oxygen we breathe.[273]

Natural resources can be categorized as follows:

Biotic natural resources come from organic matter or living things. Think fish, forests, land animals, fossil fuels. Many biotic resources can regenerate over time. For all practical purposes, however, fossil fuels cannot.

Abiotic resources come from non-organic material. Think water, metals, land.

Renewable resources such as solar, wind, and hydroelectric energy are not diminished by use.

Nonrenewable resources such as metals and fossil fuels are depleted by extraction from finite reserves.

Michele Ruta and Anthony J. Venables tell us that, "Natural resources account for 20 percent of world trade and dominate the exports of many countries."[274] Yet currently the health and abundance of those resources are not reflected in most economic assessments.[275]

To right this imbalance, our GNR index needs to encompass:

— Viability and biodiversity of essential ecosystems worldwide

— State of rivers, lakes, and aquifers

— State of fossil fuels and minerals mined from the Earth

"But, whoa! Wait one minute there!" I hear you say. "How can we capsulize this mind-boggling trove of data into a single number, an index, that's easy to grasp and understand? How can we even put a value on natural resources as diverse as coltan, anchovies, mangrove forests, and minerals on the moon?"[276]

Good point. Nature doesn't put price tags on the resources and services she delivers. Land, for instance— folks have been arguing about how to value land since biblical days, if not before. Daly and Cobb argue that for all practical purposes many economists substantially undervalue, indeed, abstract away, the value of land in their models.[277]

Hashing data into an index is no walk in the park. Economists and stakeholders

Figure 13.1: Groundhogs on alert
Credit: Wobbyball, CC BY-SA 3.0

spent decades developing the accounting systems that inform GDP, years deliberating, testing, and refining which measures to count, how to weight them, and how to combine them into a single index. Even then, politicians and special interests put their fingers on the scale.

Even though we may not know how GDP is derived, we know that it's important and likely support political leaders committed to making it grow ever bigger and ever faster.

GGH is an even bigger challenge than GDP. Envisioning an index that captures the state of both ecosystems and human welfare at global scale is a truly formidable challenge. But breathe easy. Researchers and experts from many fields supported by academic, governmental, and transnational institutions are hard at work defining both natural resource and human welfare indices.[278] As noted in Chapter 7, SEEA CF provides a global framework for auditing environmental state of affairs.

Economics professor Eli Fenichel led a 27-agency team to integrate natural capital data into US decision making. He explains that:[278]

> The National Strategy to Develop Statistics for Environmental-Economic Decisions provides a road map for accounting natural capital. It adds environmental data to the US federal balance sheet with the goal of developing relevant

statistics for decision-making.

The urgency of measuring and monitoring natural resources is well illustrated by the history of the Northwest Atlantic cod fisheries, which stretch across an 11,000 square-mile area of sea floor between Cape Cod, Massachusetts, and Cape Sable Island, Nova Scotia.

The first Europeans to sail these waters claimed that codfish were so abundant that they could scoop fish out of the water in baskets. To my mind, baskets are not a metric to spur confidence. Nevertheless, the George's Bank cod fishery off Cape Cod was New England's first industry and for long the cornerstone of the New England economy. Between 1768 and 1772, codfish accounted for 35 percent of New England exports overseas.[279] The Grand Banks off Newfoundland supported at peak years more than 35,000 fishermen and plant workers from 400 coastal communities.

But in 1992, Northern cod populations fell to one percent of historic levels due to overfishing, with harsh effects on New England and the Canadian Maritimes.[280]

During the first half of the twentieth century the annual codfish catch off Georges Bank ranged between 200 and 300 tons. But it more than doubled in the 1950s when fleets of European, Soviet, and Japanese factory ships moved in. By the late 1960s some 133 foreign fishing vessels were trawling the cod-spawning grounds, scooping up fish before they could reproduce. These vessels dwarfed the local fishing fleets in both vessel size and capacity. The enormous nets towed by the trawlers could ensnare 200 tons (181.4 metric tons) of fish an hour. At peak year 1968, the gigantic trawlers harvested more than 810,000 tons (734,827 metric tons) of cod.[281]

Failure to prudently manage the Georges Bank cod fishery imposed long-term consequences. Overfishing plus damage to the rocky bottom and ground fish ecosystem of Georges Bank decimated the stock of codfish and overwhelmed the population's reproductive capacity. Damage to the rocky bottom and ecosystem was caused by heavy chains that scraped across the rocks to hold down trawling nets.

Even with tens of thousands of jobs and millions of dollars at stake, the unsustainable rise of annual catch was not a sufficiently compelling metric to inspire sustainable management of the codfish population on Georges Bank.

Factory ships plundered the Georges Bank fishery for more than a quarter of a century, until 1976, when the US enacted the Magnuson-Stevens Fishery Conservation and Management Act. The Act made it illegal for foreign fishing vessels to fish inside a 200-mile limit from US shoreline and imposed regulations to rebuild dwindling fish stocks.[282]

Canada imposed a 200-mile limit on the Grand Banks fishery in 1977. Then, in 1992, it banned cod fishing along its east coast. The moratorium ended nearly five centuries of cod fishing in Newfoundland and Labrador. Canada didn't reopen the fishery until 2024.[283]

But, even with oversight and regulation, the codfish industry has failed to rebound. In 2021 the New England Fishery Management Council called for a reduction of the commercial Georges Bank cod catch from about 2.4 million pounds to 540,000 pounds.[284]

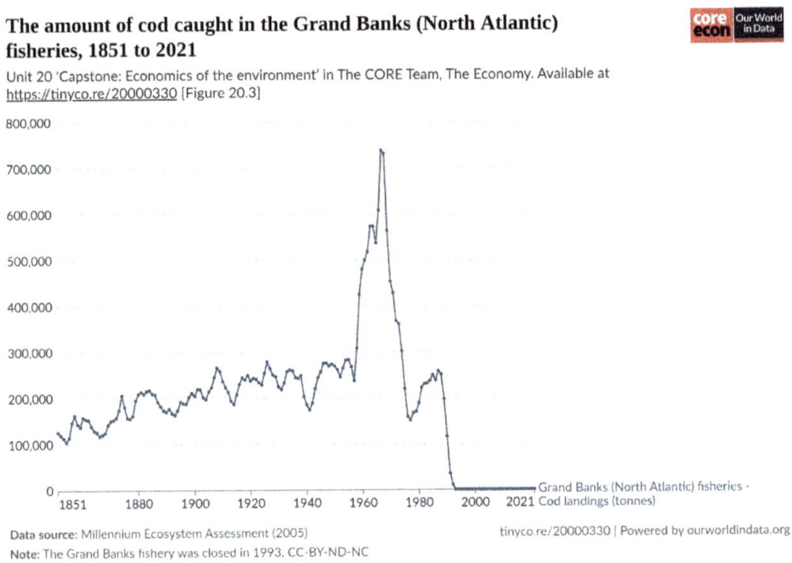

Figure 13.2: Cod caught off the Grand Banks
Credit: ourworldindata.org CC-BY-ND-NC

Overfishing isn't the only threat to codfish populations. Climate change is taking a toll:[285]

> Recent 'ocean heat waves' have occurred in the Gulf of Maine, which is warming faster than 99 percent of the world's oceans and is beginning to lose its subarctic characteristics. Ocean warming, acidification, and sea level rise are already affecting ecosystems and coastal areas, and will continue to accelerate as the climate warms.

Now you're wondering— can codfish stock off the Northwest Atlantic coast ever be replenished?

Hope springs eternal.

In 2003 the Massachusetts Division of Marine Fisheries initiated the Gulf of Maine Atlantic Cod Industry-Based Survey to take stock of the remaining codfish population. Over a five-year period the survey intensively sampled the core area for Gulf of Maine cod. The survey involved more than 3,500 survey cruises using four commercial vessels fishing standardized bottom trawls. Each cruise collected biological data, such as length, age, sex, and maturity, of more than 100,000 individual cod. Findings have helped improve management of the cod fishery in the Gulf of Maine.[286]

In 2023 the regulatory New England Fishery Management Council approved a strategy rebuilding the stock by 2033. The Council predicted a 70 percent chance of success.[287]

There's a lesson here: it's never too late to monitor and manage essential ecosystems, even if it means curtailing production to allow nature to heal itself.

Imagine now every resource on Earth that we extract from nature— everything from alfalfa to zirconium. Think of the impact on lives and livelihoods should we manage them as imprudently as the pre-1970s Northwest Atlantic cod fisheries.

The CIA *World Fact Book* lists primary natural resources by country.[288] The *World Bank Commodities Price Data (The Pink Sheet)* posts price data.[289] And the *World Atlas* ranks countries by estimated value of natural resources.[290]

Fossil fuels account for much of the natural resource wealth.[291]

Environmentalist Bill McKibben points out that fossil fuels are

> ...only available in a few places around the world...that means that the people who control those small and scattered deposits have way too much wealth and power—which they have used to dominate the rest of us.
>
> By contrast, the sun is everywhere, a liberating force precisely because it can't be hoarded or held in reserves.[292]

Our challenge is to move beyond the narrow interests of those who hoard and control natural resources toward the greater good of equitable distribution and long-term sustainability of natural resources. In a doughnut world, water, arable land, and harmonious human relations would likely be far more valuable than, say, fossil fuels and uranium.

Findings from the UN's 2024 Global Resources Outlook highlight the urgency of environmental assessments and point the way:[293]

— Sustainable and equitable natural resource use and management are essential.

— Increasing resource use is the main driver of our planetary crisis.

— Material use has increased more than three times over the past 50 years.

— Climate and biodiversity impacts from materials extraction and processing greatly exceed targets based on staying within 1.5°C of climate change and avoiding biodiversity loss.

— Environmental impacts of resource use need to fall while well-being contributions from resource use rise.

— High-income countries use six times more materials per capita and are responsible for ten times more climate impacts per capita than low-income countries.

— Compared to historic trends, it is possible to reduce resource use while growing the economy and improving well-being.

— Bold policy action is critical to phase out unsustainable activities, speed up responsible and innovative ways of meeting human needs, and promote social acceptance of the necessary transitions.

— The prevailing approach of focusing almost exclusively on supply (production) must be supplemented with a much stronger focus on demand-side (consumption) measures.

— The scientific community is united around the urgency of resolute action and bold decisions that protect the interests and well-being of all, including future generations.

Some may deride these findings as advocacy of degrowth— Phil Mullan, contributor to *Spiked* magazine, calls degrowth "...a leftish environmentalist, economic argument against 'growth-obsessed' capitalism." Calls for degrowth, he says, "represent a rejection of modernity in its totality."[294]

Since when was growth of wisdom a rejection of modernity?

But opponents of degrowth have a point. For our GGH proposal to take root, it must be adopted, reported, and maintained comprehensively around the world. Quite reasonably, many folks in poor countries who've seen their natural resources exploited by others will scoff— "You've exploited our resources in the name of globalization, but now you want us to protect what's left. You tell us that we should not pursue economic growth." What to do?

What can we do?

For one, we need to listen. Every person on Earth deserves a seat at the table. Every perspective deserves thoughtful consideration. Nature is at the head of the table. We need a translator. Metrics enable us to hear nature's urgent pleas—harvest what we need; heed when we exceed her reproductive capacity; restore when we've taken too much. Our goal is abundance for all now and sustainably into the future. We need the wisest minds on Earth and certified experts to help us take stock.

This brings us to Global Human Welfare. Sustainability of essential natural resources is essential to human welfare. Our lives and livelihoods depend upon a multitude of resources that we harvest from the natural world.

From a 21st-century perspective, I'd argue that the following conditions are essential elements of human welfare:

— Clean air

— Potable water

— Nutritious food

— Safe and comfortable shelter

— Good health and healthcare

— Domestic and national security

— Convenient transport

— Universal education

— Vibrant communities

— Accountable governance

The UN's Department of Economic and Social Welfare's 17 Sustainable Development Goals address many of the issues we'd include in our GHW index. In 2023 the UN published a chart that provides a visual framework of progress toward each of the 17 goals, based upon trend data reported between 2015 and 2022.[95] The analysts had sufficient data to assess 138 of 169 indicators. Findings:

> ...a mere 15 percent are on track to be achieved by 2030. Nearly half — 48 percent — of the targets that can be assessed show moderate or severe deviations from the desired trajectory.

It's clear that much work is yet to be done to define and deploy our global indicators. But to continue in the spirit of Don Quixote, let's give thought to design of an eye-catching, thought-provoking, and emotionally compelling icon — a dashboard readout of GGH — based on Kate Raworth's doughnut visual framework.

My take: we need a super-talented graphic designer. Prime design spec: One glance should tell the public, policymakers, and activists where threats to habitability are chipping away at quality of life on Earth.

Our icon must reflect sustainability metrics across four spheres of concern:

— *Atmosphere:* the composition of carbon dioxide, methane, nitrous oxide, and water vapor that regulate temperature of the Earth and sustains all terrestrial life

— *Biosphere:* resources drawn from regenerative ecosystems

— *Lithosphere:* mineral resources available for extraction and recycling

— *Econosphere:* health of economic and political systems and institutions that serve human needs and wants

We do indeed have our work cut out for us.

14

Humanity 2.0

Imagine a totally automated non-profit global enterprise with a storehouse of water, minerals, and organic materials that has churned away without human intervention for 4.5 billion years, recycling carbon dioxide for plants, oxygen for animals, nutrients and water for all.

Nature. Ever evolving.

Nature draws energy from the sun and operates as a totally circular economy based on free trade of energy, minerals, and organic resources.

Stand back and marvel at the scope and efficiency of operations; the many faceted beauty of nature's near-endless inventory of microbiota, invertebrates, plants, and animals ranging from microscopic viruses to massive blue whales.

Now consider— all global operations of nature were running without a hitch but for the occasional volcano eruption, meteor strike, and species enhancement until one of its marvelous hominid lines, more specifically, certain aggressive *Homo sapiens*, staged a takeover, declaring dominion over all and the right to harvest and desecrate global ecosystems without limit.

This cleverly assertive subspecies looted, polluted, and flourished. Cohorts within the faction contested ownership-in-common and prosecuted ever more bloody battles and wars among their kind to seize and secure exclusive proprietary ownership of nature's vast dominions and treasures.

Heady in conquest, the avaricious *Homo sapiens* propagated exponentially and exploited ever more swaths of nature. They totally extinguished many species of plants and animals, rendered waters and lands toxic, disrupted the delicate balance of gases in the atmosphere, and prosecuted ever more devastating wars

Figure 14.1: Hummingbird
Credit: CC0 1.0 UNIVERSAL

until it became evident to enlightened ones that their kind was committing ecocide— propelling their own hominid species and others toward extinction.

So here we are.

Let's take stock.

Appendix A, Earth in Real Time, lists websites that report up-to-date estimates of humanity's environmental footprint. On this day, October 6, 2024, as of 6:10 p.m. EST, I find the following:

Human Population 8,134,972,665 and trending up.

Natural Resources Since start of this day, some 185,795,678 tons of natural resources have been extracted from Earth.

Global Waste Some 4,521,750 tons of waste dumped into the environment.

Global Resource Consumption Humanity is using up natural resources, such as food, water, and land, 1.8 times faster than the planet can regenerate.

Global Atmospheric CO_2 Near record high—419.3 parts per million.

Global Average Temperature On track to be warmest year since records began.

Global Inequality Poorest half of the global population owns just 2 percent of global wealth; richest 10 percent owns 76 percent.

Doomsday Clock 89 seconds to midnight— a moment of historic danger.

Consider: Satisfaction of human needs and wants is both directly and indirectly dependent upon favorable climate, healthy ecosystems, and natural resources. Climate determines what and where we can gather or grow food. The health of ecosystems governs the reproductive capacity of natural resources. From a materialistic point of view, think of natural resources as capital accounts held in reserve by nature for benefit of humanity.

What to do?

At this moment, as I gather thoughts, Floridians are bumper-to-bumper facing gas shortages and gridlocked roads, seeking to avoid the ravages of Hurricane Milton. "If you choose to stay, you are going to die," said Tampa, Florida, mayor Jane Castor. A direct hit on Tampa, population exceeding 400,000, will no doubt result in lives lost, multi-billion dollar property damage, and serious US supply chain disruptions.

"Milton has the potential to be one of the most destructive hurricanes on record for west-central Florida," warns the National Hurricane Center.

Hurricane Milton will be the second devastating storm to wallop Florida within two weeks. Helene swept across the Southeast just 12 days ago, costing more than 230 lives and destroying countless homes and businesses with estimated property damages ranging up to $47 billion.

Now, days later, I sit down with relief. Tampa was spared a direct hit.

Ecosystem scientist William J. Ripple and colleagues document 16 extreme weather events around the world within the past year, not counting Milton and Helene.

> We are on the brink of an irreversible climate disaster. This is a global emergency beyond any doubt. Much of the very fabric of life on Earth is imperiled. We are stepping into a critical and unpredictable new phase of the climate crisis.[296]

What can we do?

Morning, October 11, 2024:

I'm watching goldfinches vector in fast as arrows to pluck a seed or two from our bird feeder, then flit off into the woods— yellow streaks vanishing into the dark and green. These tiny birds are bulking up for long migration to southern climes.

I'm thinking about the first responders to Hurricane Helene, selflessly swarming into devastation to rescue the stranded, tend to the injured, recover the dead, clear shattered debris and uprooted trees from obstructed roads, repair downed power lines, and bring food, water, and solace to the afflicted.

My wife's friend Kathy is a psychologist who helps first responders process trauma they've experienced on their jobs.

I can only imagine the emotional impact on first responders and the trauma of families who've lost all to hurricane ferocity.

I'm thinking of others on the Florida west-central coast stacking sandbags, boarding up windows, repositioning equipment and supplies, checking disaster recovery gear, directing evacuation traffic, as they prepared for Hurricane Milton.

But then, I read this headline in today's *The Guardian*: "US meteorologists face death threats as hurricane conspiracies surge."[297]

Hurricane conspiracies? Death threats? What's wrong with this picture?

Climate scientists tell us that hurricanes of Helene and Milton scale are twice as likely to happen due to global heating. And their models imply that it could get much worse. And it has. On October 2025 Hurricane Melissa hit Jamaica with devastating force.[298] What to do?

What can we do?

It occurs to me that we must all become second responders— third and fourth responders.

First step, we must introspect. Ask *why*?

— Why must we accept extreme weather events as the new normal?

— Why must we continue to plunder and pollute our planet?

— Why must we thoughtlessly heed the siren song of advertising and market hucksters to want more and more, indeed, need more and more?

— Why must we allow corporate campaigners and cynical politicians to provoke our lizard brains, stoke fear, anger, and hate within us to impede constructive climate policy and action?

Next step, we must face up to willful ignorance, greed, fanaticism, and reckless hunger for status, power, and dominance. And more, we must understand and minimize the psychological and social conditions that amplify them, such as lack of trust, social isolation, fear, and wealth and power inequalities. Psychologists and sociologists have studied these issues. Addressing them has long been the role of spiritual and wisdom traditions. The founding documents of US governance address many of these concerns through the separation of powers.

Finally, we must rethink our standing in the natural world. Over the past two centuries many have lost day-to-day connection with nature. And now, technology has lured us into cyber bubbles with loss of nurturing communities. We have become economically and politically polarized. All too many have embraced short-term, individualistic thinking and behavior.

Environmentalist Bill McKibben tells us that:

> The dystopian future is not just about the endless fires and floods; it's also about a society that pulls apart in their face, where people can't work together because they've been so divided by disinformation and hate.[299]

It's long been known by indigenous peoples living close to the land that natural resources are the rock-bottom foundation of our well-being.[300] We must now relearn how to live in balance with nature, all in a race against time.

Denial and despair are not options.

In my humble view, humanity as a whole, including governments, NGOs, transnational corporations, and individual advocates and activists, must engage on many fronts:

Global Population

- Reduce the global average number of children per family
- Coordinate and cooperate at global scale
- Sustainably secure basic needs for all

Greenhouse Gases

- Curtail fossil fuel consumption
- Reduce agricultural CO_2, NO_2, and methane emissions
- Restore degraded land
- Plant a diverse variety of trees

Waste Products

- Reduce food and other waste
- Curtail disposal of toxic waste into natural ecosystems
- Curtail nitrogen run-off from farmlands
- Curtail disposal of plastics into rivers and seas
- Research, develop, and roll out more efficient recycling technologies
- Create circular no-waste economies

Renewable Resources

- Establish sustainable natural resource limits
- Extract and consume natural resources within limits
- Restore degraded natural ecosystems

Non-renewable Resources

- Research, develop, and implement less eco-destructive and more yield-efficient resource extraction and processing technologies
- Recycle all man-made goods
- Find functional substitutes for non-renewable resources

Governance

- Support transnational efforts to develop effective frameworks for management of natural resources
- Grant legal personhood to natural ecosystems such as forests, rivers, and species at global scale
- Promote transparency of regulations governing rights to extract natural resources
- Promote incentives for sustainable extraction, management, and consumption of natural resources
- Promote equitable and just distribution of products and profits derived from nature

The challenges and obstacles are daunting. Nobody said it would be easy.

Countless individuals and organizations are facing up to the challenges. See Appendix C. Many have dedicated their lives to healing injuries we've inflicted on our planet and on those less fortunate around the globe.

Turns out, we each have more agency than we ever imagined.

Here are a few things we can learn and do:

Learn

- Review the UN's 17 Sustainability Goals. See Appendix B.
- Review and reflect on the UN's Global Framework for Managing Nature Through 2030: 1st Detailed Draft Agreement. See Appendix B.
- Review the Kunming-Montreal Global Biodiversity Framework. See Appendix B.
- Review Chapter 5, Part 2, Item 69 of the Constitution of Kenya. See Appendix B.
- Read publications that cover environmental and economic sustainability issues. See Appendix D to get started.
- Read books about climate change and environmental sustainability. You'll find a list of books consulted for this book in Appendix E.
- Evaluate your community's extreme-weather preparedness plans.
- Evaluate your community's recycling operations.
- Identify the major sources of environmental pollution in your community.
- Identify major distressed ecosystems in your community.
- Explore parks, conservation land, and nature reserves within your regional areas and beyond.
- Learn how industries in your region are addressing environmental sustainability issues.
- Learn how to effectively organize community action.

Ask

- How big is our carbon footprint? Find out in Appendix A, Earth in Real Time.
- How big is our environmental footprint? See Appendix A, Earth in Real Time.

- How big is our plastic footprint? See Appendix A, Earth in Real Time.
- How big is our water footprint? Bring up the website in Appendix A, Earth in Real Time.

Observe

- How much space does your community provide for parks and natural areas?
- How much space is devoted to cars— roads, parking spaces, and parking lots?
- How walkable is your community?
- How much time do you spend each week commuting to work or school?
- How convenient are your mass transit options?
- How much shade and tree cover does your community provide across all neighborhoods?
- How clean are local streams, rivers, and lakes in your area?
- How are your local schools addressing climate, environmental, and economic sustainability issues?

Practice

- Express respect and compassion for all citizens of Earth.
- Purchase reliable and repairable hard goods and electronics.
- Shun fast fashion.
- Buy secondhand.
- Cultivate awareness of natural beauty in your daily life.

Consume

- Live and consume within sustainable natural limits.
- Consume less red meat.
- Replace fossil fuels with clean energy.
- Replace plastic containers with glass.

Innovate

- Create ways to reduce your environmental footprint— CO_2, plastics, air and water pollution
- Create ways to reduce waste in the workplace
- Create ways to support a global circular economy
- Create ways to help sustain natural resources globally
- Use your design/maker/software development skills to make natural resource monitoring and management more transparent and efficient

Support

- Support environmentally responsible corporations
- Support organizations working toward sustainable ecosystem and species protection
- Support political candidates and office holders with strong environmental platforms
- Support local farmers' markets
- Support community gardens
- Support friends and neighbors in need
- Support animal rescue operations

Advocate

- Advocate for strong climate stabilization and mitigation legislation and policy
- Advocate against climate-change denial and despair
- Advocate for sustainable ecosystem and species protection
- Advocate for evidence-based facts and civility in public affairs
- Advocate for transparency and accountability in transnational environmental and natural resource treaties, agreements, and trade policies
- Advocate for programs and policies that meet minimum basic needs for all
- Advocate to reduce hunger and homelessness
- Launch an environmental/economic sustainability website, blog, podcast, YouTube, or TikTok channel

Mobilize

- Organize an environmental and economic sustainability book club
- Organize litter clean-up days
- Organize a neighborhood swap meet
- Organize a neighborhood tool-and-garden sharing cooperative

Institutionalize

- Work to pass environmental protection and economic sustainability legislation at all levels of government
- Work to create transparent and equitable international trade agreements
- Work to implement and support fact- and action-based environmental and economic sustainability curricula in the schools

Clearly, opportunities to make a difference abound. And this list barely touches the surface. No doubt you will think of much more to do.

The challenges before us are global. They demand global cooperation, hard self-reflection, and political work to wrestle down to human scale. Regrettably, climate-change denial, short-term natural resource exploitation, and geopolitical conflict and war obstruct obvious solutions.

We must expect aggressive push back as we become ever more effective squeaky wheels for environmental and economic sustainability. We're asking much of our global community. We're up against ignorance, fear of change, and powerful vested interests. Family and friends will argue. Trolls will propagate lies and threats across social media. But science-based facts, attentive listening, respectful diplomacy, and strategic persistence will arm us well.

Before we leap into the future, however, we must think hard about how we fit into the cycles and evolution of the natural world. And more, we must reflect with all humility on human nature.

For all its beauty and grandeur, nature is a fiercely competitive realm. Diverse organisms compete for space, nourishment, and sexual favor. Food webs are vast domains of ruthless predation and violence. British sociologist and philosopher Herbert Spencer interpreted Charles Darwin's theory of evolution as "survival of the fittest."[301] In quest of a perfect society, eugenicists advocated weeding out the "unfit" as means to improve the "human stock." Hitler forged these ideas into public policy with outcomes we know all too well.[302]

Going forward in the world, then, what is our stance? If fierce competition, predation, and violence are inextricably ingrained in our human nature, how can we possibly agree globally to live within the limits of natural resources? Is it indeed possible for us, in this age of nuclear weapons, to find within ourselves a higher nature— respect for all of nature's bounty and willingness to share with strangers? Alfred Lord Tennyson struggled with this question in his poem "In Memoriam," where he describes nature as "red in tooth and claw" in contrast with our embrace of "immortal Love." Of God, he writes,[303]

> Thou madest Death; and lo, thy foot
> Is on the skull which thou hast made.

We do know this: Humanity is powerful enough to dominate the natural world and decimate long-evolved ecosystems. We have seen envy, greed, and pursuit of power contribute to suffering, starvation, and death of millions. We have developed weapons capable of incinerating cities. We have erected walls of mistrust and hostility. We have subverted, invaded, exploited, and oppressed people who dare defy our ambitions and will. And we continue to spew waste and toxins into environments that sustain us.

But we also know that "red in tooth and claw" is not the whole story. Ecologists now recognize that cooperation is a foundation of many interactions in nature.[304] And, of course, we humans wouldn't have pot luck dinners or skyscrapers, moon landings or Little League teams, without it.

So the choice before us is this: Shall we deploy our best intelligence and wisdom to regulate the dark side of our human nature, work in balance and harmony with nature, and lift the foot of death off humanity's fragile skull or, shall we continue on our heedless way? Shall we learn to cooperate in the global management and distribution of natural resources?

My bet is that most of us would choose a path toward life, well-being, and long-term sustainability in a vibrant natural world. What to do?

What can we do?

The simple answer— there is no magic bullet. We can strive to understand how our consumer choices contribute to global warming and environmental degradation. Armed with understanding, we can make more sustainable choices. But as climate scientist Katharine Hayhoe suggests, we can do more:[305]

> ...if we want climate action to grow, we can't assume extreme

> events will speak for themselves. People need trusted messengers — neighbors, friends, colleagues, health professionals — who can explain how climate change is affecting our lives here and now. In other words, they need you!

We can become advocates and ambassadors for all living things on Earth.

Part II

Hope Springs Eternal

Those who contemplate the beauty of the earth find reserves of strength that will endure as long as life lasts.

Rachel Carson

15

Nature

The word "nature" evokes varied mental images and connotations.

The following biblical passage suggests that man is distinct from nature with proprietary right to dominate:

> And God said, Let us make man in our image, after our likeness: and let them have dominion over the fish of the sea, and over the fowl of the air, and over the cattle, and over all the Earth, and over every creeping thing that creepeth upon the earth— Genesis 1:26 (KJV)

Many indigenous peoples of North and South America, Australia, and elsewhere see nature as a source of identity, spirituality, and well-being. They express their respect through rituals, responsibilities, and duties. In Ecuador, the Shuar, for instance, have drawn on their cultural traditions, knowledge of plant medicines, and low-impact agricultural practices to develop a conservation strategy to protect their Amazonian ecosystems from loss due to mining and deforestation.[306]

For some, nature is alien territory of beasts and mosquitos. Best avoided. Is it far fetched to suggest that many contemporary urban denizens believe that nature is a frequent-flyer destination?

What do you think?

Personally, beyond the experiential wonders and beauty of "the great outdoors," I think of the many and varied life-forms and multi-faceted flows of energy that interconnect all living things— food webs; competitive, symbiotic, parasitic, and altruistic relationships between and among plants and animals; plus the atmosphere, lithosphere, and aquatic environments that host life on our planet.

Figure 15.1: Mother bird
Credit: Dikyedarling, CC BY-SA 4.0

The diversity of life from bacteria and fungi to elephants and whales is staggering. Some 100,000 plankton species abound in the sea— some nearly microscopic and others up to 13 feet long.[307]

Plankton have helped shape the mountains of Earth as their skeletal remains, along with others, accumulated at the bottom of ancient seas and were compacted under great pressure, forming limestone. The deep sedimentary layers of limestone were then pushed up to great heights by tectonic forces.[308]

Living organisms have evolved to thrive in the most extreme environments of Earth— nearly two miles under the sea floor,[309] deep under arctic ice,[310] in Yellowstone hot springs,[311] and even in microwave ovens.[312].

Did you know that mushrooms can drive robots? Cornell researchers harnessed the innate electrical signals of fungal mycelia, common on the forest floor, to control two "biohybrid" robots.[313]

My take: try as we might to put *Homo sapiens* into one distinct and separate category and nature into another— it just won't work. We're simply apex predators preening atop complex adaptive food pyramids. Like snails in spiral shells, we wrap ourselves in the structural fruits of our technologies. But nature provides the food we eat and the oxygen we breathe; consumes the CO_2 we exhale and the excreta we discharge from our bodies. The cells in our bodies share structural and functional similarities with myriad life forms embedded across the interwoven energy networks of nature. And, for every cell in our bodies, we symbiotically host countless millions of bacteria, not to mention myriad fungi and viruses— our microbiome, the multitude of independent natural life-forms dwelling in and among our cells.[314]

And more, nature provides the raw material and energy that make our wondrous technologies possible. Take away nature and we cease to exist. Remove us and nature evolves ever on. There's no getting around it. We're embedded in nature and nature is embedded in us.

Skeptical?

As you snuggle down to sleep tonight, take a deep breath. One in five molecules entering your lungs is oxygen, O_2, that may well have been emitted by tree leaves in a distant forest or by phytoplankton drifting on ocean currents half a world away. As the oxygen traverses your blood stream it is transported into cells where it reacts with nutrients from your latest meal to generate energy that powers your brain, muscles, and every organ. Now exhale. Mitochondria in your cells cast off carbon dioxide, CO_2, that diffuses out into your blood stream, on into your lungs, and out through your breath. These carbon dioxide molecules may be carried by wayward winds and ocean breezes to a distant forest or ocean waves to be sequestered by tree leaves or phytoplankton. Transformed by the magic of photosynthesis into vital nutrients, they could one day end up on your dinner plate as a lettuce leaf or tasty salmon steak. Relax now and feel the gentle pull of gravity securing you snug in your bed. That's the mass of the Earth, our home in the universe, exerting a force that keeps you from drifting out into orbit around the sun.

My conception of nature is congruent with Kate Raworth's "network of flows."[315]

Raworth stresses that:

> Resources such as energy, matter and information can flow through these networks in ways that achieve a fine balance between *efficiency* and *resilience*.[76]

You'll know this from Biology 101— energy from the sun fuels photosynthesis, a chemical reaction that transforms water, carbon dioxide in the air, and minerals from the soil into plant tissue and oxygen. Animals breathe in oxygen and eat plants to obtain the energy they need to grow and function. Predators eat prey to obtain the energy they need. At each step, energy is transformed, until it's dissipated as heat.[316] Nature has evolved globe-spanning ecosystems for billions of years to optimize the efficiency and resilience of this food chain.

The flow of energy through the network of all living things on Earth is complex and dynamic, ceaselessly adjusting and adapting to climatic conditions and stress. Some 300,000 to 1 million years ago our kind, *Homo sapiens*, evolved as but one of many interdependent connections in this flow. Throughout our history our survival has depended upon oxygen and nutrients provided by the forests, grasslands, oceans, and atmosphere of the natural world. Nothing of substance has changed over the eons other than our willful exploitation of long-evolved ecosystems and our profligate discharge of waste products, some quite toxic, into the soils, aquifers, rivers, oceans, and atmosphere of our planet. Nature will continue to evolve with or without us. But our damn-the-cost obsession with material growth drives nature to extreme disequilibrium that threatens human life on Earth. The question is, how much longer will she continue to support our lives, livelihoods, and strivings?

In short, nature could essentially edit us out of living organisms on Earth. And we'd have no one to blame but ourselves.

To grasp this in mercenary terms— globally, land and marine ecosystems provide services worth an estimated $125 trillion USD. Some 1.2 billion people work in related industries such as farming, fishing, forestry, and tourism. By 2050, however, up to 5 billion people may suffer lack of fresh water and inadequate nutrition due to diminishing ecosystem services, particularly in Africa and South Asia.[317]

Fortunately, robust efforts small and large are underway worldwide to enact policies and promote behaviors that preserve and restore ecosystem resilience.

As one example of a small project with big impact, consider the New Alchemy Institute, established by scientist John Todd; his wife, writer and activist Nancy Jack Todd; and fish biologist William McLarney. Over a two-decade period New Alchemy designed and implemented natural systems to treat wastewater and remediate degraded water bodies.[318] Their systems integrated sunlight, microorganisms, aquatic plants, and sand.

Dr. Todd founded John Todd Ecological Design in 1988 to commercialize New Alchemy-designed, nature-inspired systems for treating sewage and industrial waste and rehabilitating water bodies ranging from sewage-contaminated urban canals to storm water ponds and natural lakes. Over a period of 27 years John Todd Ecological Design constructed systems to purify water on five continents and influenced how many think about wastewater treatment and disposal.[319] Todd's book *Healing Earth: An Ecologist's Journey of Innovation and Environmental Stewardship* is an inspiring overview of his vision and accomplishments.[320]

At transnational scale, Africa's Great Green Wall was launched in 2007 to restore savannas, grasslands, and farmlands across the Sahel-Sahara region, which includes Djibouti, Eritrea, Ethiopia, Sudan, Chad, Niger, Nigeria, Mali, Burkina Faso, Mauritania, and Senegal.[321] It is one of the world's most ambitious ecosystem restoration projects.[322]

In 2021 the UN launched the Decade on Ecosystem Restoration. The initiative calls for "the protection and revival of ecosystems all around the world."[323] Major projects are underway in Brazil, Paraguay, Argentina, United Arab Emirates, India, Serbia, Kyrgyzstan, Uganda, Rwanda, Costa Rica, El Salvador, Guatemala, Honduras, Nicaragua, Panama, Indonesia, China, Vanuatu, St. Lucia, and Comoros.

Climate adaptation and ecosystem restoration pay big dividends.

Will Anderson, writing in 2021 for the World Resources Institute, tells us that ecosystem restoration in the US is a $25 billion USD industry and employs some 220,000 people. He further reports that each $1 invested in restoring degraded landscapes can bring economic returns of as much as $7 to $30.[324] A 2025 report from the World Resources Institute (WRI) found much the same. It evaluated 320 climate adaptation and resilience investments across 12 countries. WRI researchers found that every $1 invested in adaptation can yield over $10.50 in benefits.[325]

Many cities around the world have initiated programs to promote biodiversity and to mitigate the urban heat-island effect.[326] Opportunities abound:

— Expand urban green spaces such as parks and urban forests

— Rehabilitate vacant lots and landfills

— Promote urban and suburban gardens

— Support greening of campus areas and public spaces

— Better regulate streams, lakes, ponds, and water runoff with artificial swales and storm retention ponds

Benefits include water regulation, pollution reduction, habitat restoration, and various health and cultural perks.

In the hustle of daily life, surrounded by asphalt, concrete, drywall, and glass, it's easy to forget our existential debt to the natural world. Human welfare is fundamentally connected to the flow of ocean currents, phytoplankton producing oxygen while feeding whales, microbes in agrarian soil, biodiverse rainforests, and wind-driven water vapor condensing in clouds and falling as rain over highlands, forests, and savannas.

Urban green spaces are modest reminders that there's more to life than shopping.

16

Water

On hot days we're advised to stay hydrated. A UN report tells us, however, that between two and three billion people globally experience water shortages for at least one month per year. Two billion people (26 percent of the global population) do not have safe drinking water, and 3.6 billion (46 percent) lack access to safely managed sanitation. The growing incidence of extreme and prolonged droughts is also stressing ecosystems, with dire consequences for both plant and animal species.[327]

We've seen hot days break temperature record upon record over the past decade. In August 2024 the global surface temperature was 2.29°F (1.27°C) above the 20th-century average of 60.1°F (15.6°C). August 2024 was the warmest August on record.[328]

Hot weather increases the rate of evaporation of surface waters. Meanwhile, drawdown for agriculture, industry, and municipal use creates ever-greater demand for water. Result: lakes are losing water and drying up around the world.

The Aral Sea, for instance, lying between Kazakhstan to its north and Uzbekistan to its south, was once one of the four largest lakes in the world, with an area of 68,000 square kilometers. But it's been shrinking steadily since 1960; now it's mostly a giant desert.[329]

Lake Colhué Huapí in south-central Argentina, which once spanned almost 800 square kilometers, roughly the size of New York City, has all but disappeared.[330]

Water level of Lake Mead in Arizona, largest reservoir in the US, is just 34 percent of capacity.[331] Lake Mead is the source of water for millions of people across seven states, tribal lands, and northern Mexico.[332]

Figure 16.1: The basin of Mexico circa 1519
Credit: Historical Free Art License

Mexico City, including suburbs, has an estimated population of 22.75 million. It is the largest, most populated metropolitan area in North America.[333]

Mexico City and surrounding metropolitan areas may soon run out of drinking water.[334] Agricultural drawdown, drought, pollution, and leaky water systems have severely stressed the water supply. Mexico City is also sinking at an alarming, perhaps unstoppable, rate of up to 50 centimeters a year due to drainage of aquifers.[335]

As you can see from the 1519 basin of Mexico map (Figure 16.3), present-day Mexico City was once a system of lakes that supported the Toltec and Aztec pre-Columbian civilizations. The Aztecs created floating gardens called chinampas on the shallow lakes of Xochimilco and Chalco some 2,000 years ago. The chinampas created tillable land for agriculture on the surface of the lakes.[336] They are considered the most productive food system in Mexico's history. But for a few canals, Lakes Xochimilco and Chalco have been drained to reduce flooding and have largely disappeared. Nevertheless, floating gardens are cultivated in the Xochimilco canals to this day to produce food, flowers, and ornamental plants. The UN considers them a Globally Important Agricultural Heritage System.[337] Some suggest that floating gardens based on chinampas could provide food for megacities of the future.

Climate scientist Fangfang Yao and colleagues analyzed water loss in what had been the world's largest lakes in 1972. Using three decades of satellite observations, climate data, and hydrologic models, they estimated that from 1992 to 2020 the lakes lost some 5.7 trillion gallons per year.[338]

In addition to water shortages in many areas, climate change makes record-breaking floods the new normal.[339] The first six months of 2023 saw nearly 100 major flood events around the world.[340]

In the first weeks of September 2024:[341]

— Floods claimed more than 1,000 lives and drove nearly 1 million people from their homes across northwest and north-central Africa.

— Storm Boris forced tens of thousands to evacuate flooded cities and caused billions in damage across Central Europe.

— Typhoon Yagi caused floods and landslides that took more than 500 lives and cost over $13 billion across the Philippines, northern Vietnam, and the highlands of Southeast Asia.

— Hurricane Helene struck the Florida Gulf Coast and moved inland across the southeast US, causing floods and landslides— $78.7 billion in damages; nearly 250 lives lost.[342]

So what's going on? Not enough water for many and far too much for others.

The World Bank Group (WBG) tells us that:[327]

— Nine out of 10 natural disasters are water-related.

— By 2050, flood and droughts could cause $5.6 trillion USD in cumulative losses to the global economy.

Water is the Earth's largest and most crucial natural resource. But only three percent of it is freshwater, and only one percent is available for use. The rest is frozen in glaciers or too deep to pump. More than 2 billion people depend upon water pumped from aquifers.[343]

The global water supply is further compromised by river and lake pollution. Rain washes nutrients in fertilizer into rivers, lakes, and coastal waters, which promotes harmful algal blooms. Researchers predict that algal blooms will increase by at least 20 percent by 2050. They further estimate that up to 80 percent of global wastewater enters untreated into rivers and lakes with harm to human health and ecosystems.[344]

Nearly three-quarters of the world's freshwater supply is used in agriculture. Industrial processes and municipalities consume the rest.[345]

Chunyang He and colleagues project daunting water-scarcity challenges for cities due to urbanization and climate change:[346]

— India will be most severely affected in terms of growth in urban populations experiencing water-scarcity (increase of 153–422 million people).

— The number of large cities exposed to water scarcity will increase from 193 to 284, including 10–20 megacities.

— More than two-thirds of water-scarce cities can relieve water scarcity by infrastructure investment, but large-scale water scarcity solutions may entail significant environmental trade-offs.

Many countries are turning to desalination to provide or supplement water needs.

The World Population Review reports that worldwide there are 569 desalination plants that are either the largest water plants in their countries or have a daily capacity of 100,000 cubic meters.[347] The Bahamas, Maldives, and Malta rely almost entirely on desalination.[348]

Current desalination technologies pose two environmental challenges, however: they require considerable energy and they discharge high-salinity brine. Fortunately, clean energy such as wind, solar, and wave power, plus in-the-lab technologies,[349] can help address energy needs. And the brine can be sustainably mined to recover critical metals.[350]

Providing sustainable water supplies worldwide will come at considerable cost. In 2023 UN Secretary-General António Guterres outlined four key areas through which to tackle the global water crisis:[351]

— Improve management of water resources worldwide.

— Massively invest in water and sanitation systems.

— Invest in disaster-resilient pipelines, water-delivery infrastructure, and wastewater treatment plants, plus find new ways to recycle and conserve water.

— Address climate change.

The WBG tells us that decades of poor management, inadequate financing, and low levels of investment in water infrastructure have imposed huge costs on society, especially the poor. Very few water or wastewater utilities in the developing world recover adequate operation and maintenance costs from customers, and only a handful recover debt service and depreciation.[352]

The World Bank and nine Multilateral Development Banks have jointly committed $50 billion USD plus technical assistance to the water sector, benefitting more than 400 million people globally.

Help can't come soon enough. As of the February 2024 United Nations World Water Development Report "None of the SDG 6 targets appear to be on track."[353] What to do?

What can we do?

The Global Commission on the Economics of Water has presented a framework for valuation, governance, and use of water based on the assumption that the movement of water around the globe is a common good.[354]

The Commission calls for collaboration and concerted action "...across sectors and scales, from local to global to address water problems worldwide." It suggests that attention, financing, and planning should shift from fixing water supply problems after the fact to shaping economies so that water is allocated and used efficiently, equitably, and sustainably from the start.

17

Forests

I sat once for a time out of time in a birch grove in the mountains of Vermont. The rhythm of white trunks gleaming, hush of rustling leaves, faint intermittent bird song, subdued the chatter of my mind. Air so fresh. I watched an ant poised on the edge of a curling leaf.

As I now sit to write about forests, the vivid image of that passing moment near 30 years ago seems like yesterday.

If you've ever strolled among the giant sequoias that live on the western slopes of California's Sierra Nevada mountain range or through an old-growth coastal redwood forest in Northern California, you know the power of trees to inspire awe and a special kind of spiritual reverence. Coastal redwoods are the tallest trees in the world. Giant sequoias are the largest trees by volume.[355]

California's coast redwood trees (*Sequoia sempervirens*) can grow from seeds no larger than tomato seeds to a height of 367 feet (112 m) and width of 22 feet (7m) at the base. Imagine a 35-story skyscraper.[356] The General Sherman giant sequoia (*Sequoiadendron giganteum*) in the Sequoia National Park is 100 feet wide at its trunk.[357] The General Sherman is estimated to be 2,200 years old.[358]

Among heights of hubris are the three redwood tree drive-through tourist attractions in Northern California.[359] Was it dreams of tourist dollars pouring into their pockets that inspired the entrepreneurs to hack car-spanning tunnels through the trunks of three magnificent *Sequoia sempervirens?* Seems to me that the thrill of driving through a tree is small beer compared with simply looking up into lofty foliage.

As you gaze up from the understory of a coastal redwood forest wet with rain

Figure 17.1: California coast redwood forest
Credit: Allie Caulfield CC BY 2.0

you might wonder: How do they survive during frequent periods of heavy fog and rain, when water limits photosynthesis?[360] If you're looking up at giant sequoias you might ask: How do the they survive periodic forest fires?[361]

It turns out that over eons redwoods have evolved super survival powers. Coastal redwoods have two types of leaves. One type converts carbon dioxide to sugars through photosynthesis; the other absorbs water. This means that coastal redwoods can thrive in fog and mist.[362] Following devastating fires that incinerate their needles, giant sequoias pour stored sugars into buds that have been lying dormant for decades, nurturing new growth.[363]

There's more here than I could ever hope to unpack in this short chapter and much more to learn about the lush ecosystems that redwood forests support.

Redwood trees have significant commercial value. Redwood lumber has beautiful hue. It's rot resistant and easy to work. For these reasons, redwood is much valued for residential decks and other outdoor structures. The National Park Service estimates that the General Sherman, the largest living redwood tree by volume, would yield about 675,000 board feet of lumber.[364] A typical redwood tree weighs some 50,000 pounds and can yield as much as 2,000-3,000 board feet of lumber.[365] Recent cost of redwood per board foot retail in the US ranges up to $12 USD.

Redwood tree fossils date back to the Jurassic period— some 200 million years ago. The magnificent trees once grew throughout the Northern Hemisphere. California coastal redwoods thrived across an area three times the size of Rhode

Island— ranging from south of Big Sur to beyond the Oregon border. But demand for lumber exploded shortly after sawmill operator James Wilson Marshall's discovery of gold at Sutter's Creek, California in 1848. Commercial logging and clearing of redwood forests became a major industry as hundreds of thousands of people rushed into the state seeking fortune. The industry has flourished ever since. Today, only five percent of the original old-growth redwood forest remains.[366]

Organizations have been working to protect the redwoods since at least 1892, when the Sierra Club was founded. One of the Sierra Club's founding goals was "to enlist the support and cooperation of the people and the government in preserving the forest and other natural resources of the Sierra Nevada Mountains."[367] Another group, Save the Redwoods League, has worked to protect, restore, and conserve redwood forests since 1918.[368]

In the 1960s, the Sierra Club, Save the Redwoods, and other conservation advocates banded together to achieve a major victory for redwood forests. This was a time of heated conflict between the lumber industry and conservationists. While speaking before the Western Wood Products Association in San Francisco on March 12, 1966, then California Governor Ronald Reagan said:

> ...if you looked at a hundred thousand acres or so of trees — you know, a tree is a tree, how many more do you need to look at?[369]

In 1968 conservationists' lobbying led to a political compromise that helped quiet the issue:[370]

— The US Congress authorized creation of the 58,000-acre Redwood National Park, which included 27,500 acres of state parkland.

— 5,000 acres of coastal property in Del Norte and Humboldt counties were ceded to the state of California by lumber companies.

— 13,000 acres of Forest Service land in Del Norte County (known as the Northern Redwood Purchase Unit) were transferred to the lumber companies.

According to Save the Redwood League's 2023-24 Annual Report, it has now helped to protect more than 220,000 acres of redwood forest.[371] The Sierra Club's forest-protection efforts continue as well. They have expanded to include establishment of a Forest Carbon Reserve that would protect older forests in the contiguous US and Canada.[372]

Despite work by these and other dedicated organizations, however, conflict between environmental activists and loggers persists.[373] And redwoods are but one of many forest species worthy of protection.

Paddle through a maze of mangrove trees and you'll experience a different tenor of awe. Leaning over the coaming of your kayak, you'll see roots plunging down into the murk like fat fingers, crabs clinging to submerged knuckles, and fish swimming in and out of dark voids. You may startle nesting seabirds who'll lift off with the sound of a thousand blankets flapping in a stiff breeze.

Healthy mangrove forests are critical ecosystems— hot spots of biodiversity that store carbon; protect juvenile fish; provide natural barriers that slow down waves and storm surges, prevent erosion, and reduce flood depths; and support livelihoods for coastal communities around the world. But the total footprint of mangroves has been declining over the past 35 years.[374]

UNESCO reports that mangroves are disappearing three-to-five times faster than global forest overall.[375]

For decades that loss was largely driven by economics— mangroves cleared to make way for shrimp farms and palm plantations. Not anymore. Global warming now poses a greater and increasing threat.

In 2024 the International Union for Conservation of Nature (IUCN) published a first-of-its-kind report on the health of mangrove ecosystems across the globe. Its conclusions reflect on-the-ground assessments from 250 mangrove experts in 44 countries. Sea level rise emerged as the primary threat to mangrove ecosystems, followed by more frequent and severe storms. The IUCN warns: "More than 50 percent of the world's mangrove ecosystems are at risk of collapse."[376]

UNESCO is acting to protect mangroves through networks of biosphere reserves, global geoparks, and natural World Heritage sites. UNESCO partner Conservation International helps to connect science and policy to ensure the protection of mangroves and coastal ecosystems. Carbon sequestered by the oceans and coastal ecosystems is known as "blue carbon." Australia, Pakistan, Colombia, Seychelles, Fiji, Indonesia, and other nations have committed to include and champion blue carbon in their national climate plans.[377]

Forests are found around the world in northern, temperate, and tropical regions. Dense stands of spruce, fir, and cedar blanket northern boreal forests, while oaks and maples in my temperate woods reward me each autumn with their vibrant display, morphing from greens to yellows to orange. Tropical rainforests, the most biodiverse on Earth, host at least 40,000 — and possibly as many as 53,000 — species of trees.[378] Mangrove tropical forests have adapted to changing levels of fresh and saltwater along brackish coastlines and tidal rivers.[379]

Some 70 million people worldwide call forests home, including many indigenous communities. More than 1.6 billion people depend on forests for food or fuel.

Forests provide oxygen, shelter, jobs, water, nourishment, and fuel. The fate of our forests may determine our own fate as well.[380]

Some 10,000 years ago 71 percent of the Earth's land surface was covered by forests, shrubs, and wild grasslands.[381] As of 2018 humanity had destroyed one third of the world's forests to expand cropland, grazing, and urban areas.[382]

Extended droughts accelerated by climate change have further stressed trees and dried the Earth to create tinderbox conditions in many forests around the world. The area burned by forest fires increased by some 5.4 percent per year over the years from 2001 to 2023. By the end of 2023 forest fires accounted for nearly 6 million more hectares of tree cover loss than they did in 2001— an area roughly the size of Croatia.[383]

Wildfires that ravaged forests in Canada and parts of Amazonia in 2023 were three times more likely due to climate change, according to a collaboration of researchers from four major United Kingdom research centers. The fires contributed substantially to global CO_2 emissions.[384] The Canadian fires alone released 647 megatons of carbon into the atmosphere— more than Germany, Japan, Russia, and four other of the prior year's ten largest emitters.[385]

It's guesswork to predict the future costs of extreme-weather events due to carbon released into the atmosphere by forest fires. But we do know that forest fires impose heavy costs on national economies year in and year out. Within the first nine months of 2023, wildfires cost Europe €4.1 billion in damages.[386] The US Joint Economic Committee estimates that wildfires in the US cost between $394 billion USD to $893 billion USD each year.[387]

Environmental costs are also difficult to project. While wildfires can be beneficial to some forests by opening seeds, the extreme heat of severe forest fires can kill viable seeds and soil microbiota, significantly slowing regrowth.[388] Essential soil microorganisms can take 25 years or more to recover from high-severity wildfires.[389]

As global heating impacts biodiverse habitats, many terrestrial organisms are shifting in latitude or elevation. According to one study, some organisms "have recently shifted to higher elevations at a median rate of 11.0 meters per decade, and to higher latitudes at a median rate of 16.9 kilometers per decade."[390]

But trees are not notably mobile. Oak trees, for instance, have thrived continuously in Britain for 12,000 years. Ecologist Charlie Gardner suggests that they may need help to move north in response to global heating.[391] Other ecologists believe that English woodlands must be made more resistant to heat waves and drought by introducing southern species. This idea, called assisted migration, is divisive. Conservationists argue that introducing non-native species meddles with nature; introduces invasive species that can damage native environments. Others say those arguments ignore the far greater risk of projected ravages from climate change.

In 2017 the UN Department of Economic and Social Affairs published *The United Nations Strategic Plan for Forests 2030*, a framework for action to sustainably manage forests and halt deforestation and forest degradation around the world. The plan spells out six goals and 26 associated targets.[392]

In 2021 110 countries signed the COP26 Glasgow Leaders Declaration on Forests and Land Use. The declaration is a commitment to reverse deforestation by 2030.[393] These nations contain 85 percent of the planet's forests.

Countries around the world have made progress toward increasing the global forest area through afforestation and restoration. Asia, Europe, Oceania, and other regions appear to be on track to reach one of the UN's key targets.[394] But on many goals, the world is falling behind.[395] Protecting and expanding forests around the world is a race against the chainsaws of loggers, bulldozers of ranchers, and the devastation of forest fires.

Costa Rica's forest management, however, gives us hope. Costa Rica is the first tropical country in the world to reverse deforestation.[396] It demonstrates what can be accomplished when political will, economic incentives, and community engagement align.

A few weeks before writing this chapter, I had the privilege and pleasure of sitting on a veranda in Sarapiqui, Costa Rica, watching rain nourish the forest and hummingbirds nourish themselves.

I watched as one bird arched its back as gracefully as a ballerina and approached a feeder just feet from where I sat. Wings flapping, white tail feathers spread, it sipped nectar through a stiletto-thin beak. Others zipped in almost faster than the eye could follow.

The rainforest beyond dazzled my eye. Brazilian red cloak flowers blazed scarlet. Cecropia trees soared. The hummingbirds perched on ferns as tall as a man. They drank from delicate heliconia flowers, drawn by the brilliant reds and oranges of the protective bracts surrounding each bud.

I marveled at the abundance of life before me. Had I been there 40 years before, however, I would have seen none of this. No forest, no birds. In 1985 this lush expanse was an abandoned palm plantation. Dave Lando, an American expat, bought the land with a dream of rewilding. He had a passion for habitat restoration, a knowledge of tropical agriculture, and a vision of what could be. He began planting native species. The forest regrew. Flowers bloomed. Birds returned. Today, Dave & Dave's Nature Park (named for Dave and his son) is home to more than 230 species of birds and an ecotourism destination.[397]

For millennia forest covered some 75 percent of Costa Rica. But logging and clearing ramped up in the 1940s. By the time Dave first plunged his shovel into the earth to plant his cecropia trees, forests covered only 21 percent of the nation.[398] Economic incentives to clear forests for pasture and farmland had taken their toll.

In the early 1990s, however, the government took action to realign incentives in favor of environmental protection. Laws made cutting down forests without permission illegal. They established national parks and encouraged individuals and NGOs to create private preserves— most notably by paying farmers and others to replant, restore, and protect forests. And they taught them how.

Result: close to 60 percent of Costa Rica is now forested.[399] Forested areas include protected national parks and private preserves that cover an astonishing 25 percent of the country.[398] Tourism has supplanted agriculture as Costa Rica's largest industry. Cloud forests in the mountains, tropical rainforests in the lowlands, and mangroves along brackish rivers are among the natural wonders that draw three million tourists a year eager to glimpse a sloth lolling high in the branches of a cecropia tree, a hummingbird sipping deep from a heliconia flower far below.[399]

My interlude on the veranda at Dave & Dave's was yet another memorable time out of time.

Feeling stressed? Suggestion— find a forest, walk in, sit, listen, take a deep breath, and look up in awe at one of nature's wonders.

18

Fisheries

I owe a blood debt to fish.

In my late teens and early twenties I was a passionate participant in the competitive sport of underwater spearfishing. I spent countless weekend hours off California shores, and not a few weekdays, diving down to depths of 70 feet or more while holding my breath hunting for ling cod, cabezon, vermilion rockfish, California halibut, and other species. I won trophies, participated in the National Underwater Spearfishing Competition off Newport, R.I., was a contender for the US National Underwater Spearfishing team. Over the course of a decade I shot hundreds of pounds of fish until one day I couldn't do it. I'd track a fish with my speargun, think "what a beauty," then let it swim on its way.

In all fairness, every fish I shot ended up on my family's dinner table; fish shot in competition were donated to homeless shelters or other charities. The number of fish harvested worldwide by spear is minuscule compared with hook-and-line shore, skiff, and commercial fisheries, not to mention bottom trawling and gill nets.

I mention my hundreds of hours stalking fish underwater only because I've experienced the cathedral beauty of kelp forests with shoals of blue, black, and olive rockfish swirling among the fronds; seen firsthand that fish are as curious about me as I of them, warily swimming in close as I lay motionless, stomach-down on the ocean floor. I've seen the over exploitation and disappearance of fish around a once-teeming rock outcropping off Cannery Row in Monterey, California, and, perhaps too far into my sordid competitive spearfishing career, came to understand that fish feel pain.

Global consumption of fish and other aquatic foods has increased more than five

Figure 18.1: Yellowtail Scad
Credit: Donvikro, CC BY-SA 4.0

times since 1960, when I was spearing ling cod in Monterey Bay.[400] Stanford University researcher Rosamond Naylor projects a further 80 percent increase by 2050.[401]

The value of world fisheries and aquaculture production reached a record $472 billion USD as of 2022, according to a 2024 report by the Food and Agriculture Organization of the United Nations (FAO). The industry employed nearly 62 million people and produced 223.3 million metric tons of fish, seafood, and algae. For the first time, most of this bounty was farmed, not fished: aquaculture delivered 130.9 million metric tons of product vs 92.3 million metric tons from wild-caught fisheries.[402] Some 40 percent of fish caught in the wild were captured by small-scale fishers in developing countries.[403]

Diversity of life in the ocean and seas supports three pillars of sustainable development:

— Economic, social, and environmental development

— Healthy functioning of the planet

— Health, well-being, and prosperity of humanity[404]

Fish provide essential micronutrients for millions around the globe. But fish populations are declining due to overfishing, climate change, pollution, and competing uses for freshwater. Marine scientists project that "...more than 10% of the global population could face micronutrient and fatty-acid deficiencies driven by fish declines over the coming decades, especially in the developing nations at the Equator."[405]

Overfishing is a serious problem globally. Based on a study of some 600 fish subspecies populations, the FAO classified more than a third (35 percent) of the world's capture fish stocks as overfished.[406]

Ironically, aquaculture contributes to overfishing, primarily of sardines[407] and anchovies.[408] Farm-raised salmon and shrimp are fed highly nutritious anchovies and sardines caught in poorly regulated ocean waters— 20 to 30 percent of the global catch. Reduction and loss of fish populations not only threaten global food supply. They also reduce marine and inland biodiversity. Michael Moyer reports in *Scientific American:*[409]

- Hammerhead sharks: population down by 89 percent since 1986
- Russian sturgeon: population down 90 percent since 1965
- Yellowmouth grouper: may exist only in pockets of its former range
- European eel: population down by 80 percent since 1968
- Orange roughy: population off the coast of New Zealand down by 80 percent

The Natural Resources Defense Council reports that: "Over one-third of marine mammals and nearly one-third of sharks, shark relatives, and reef-forming corals are threatened with extinction."[410]

Climate change is a further threat. The oceans absorb some 90 percent of excess heat and about a quarter of the carbon dioxide from fossil fuel combustion. This increases acidification and reduces oxygen levels. Seashells fail to develop properly; algae blooms choke coastal fisheries.[411]

As of 2024 temperatures in Australia's Great Barrier Reef were the hottest in at least 400 years.[412] During summer 2023 the surface ocean temperature around the Florida Keys, US, soared to 101.19°F (38.43°C).[413]

The Keys weren't the only area heating up. In 2023 nearly all of Earth 's oceans—96 percent—experienced heatwaves that lasted four times longer than the historical average. Researchers identified different regional causes for these "super-marine heatwaves," but warned that they "...could signal a tipping point in Earth's climate system."[414] Indeed, sea surface temperatures reached new highs in 2024.[415]

Another concern: microplastic pollution.

The annual production of plastics worldwide reached 460 million metric tons in 2024.[402] Plastics contain more than 10,000 chemicals, including carcinogens and endocrine disruptors.[416]

Microplastics are plastic particles less than five mm (0.20 in.). Some microplastics are generated during manufacture of plastic-based products such as synthetic fabrics; others form as wind and waves grind up plastic debris.

These tiny particles are now found in marine waters all around the world.[417] Researchers collected water samples and 240 fish, squid, and shrimp specimens across the tropical eastern Pacific and the Galápagos archipelago. Microplastic particles were found in 100 percent of the water samples and marine organisms.[418] Microplastics have even been found in the breath of wild dolphins.[418]

Not enough is known yet about the threat to marine life and humans, but research suggests microplastics may damage DNA, change how genes work, and contribute to inflammation—all known risks for disease. The plastics are found in every human organ, including the brain. They're in saliva and blood, placenta and breast milk.[419] And of course, they're in the fish we eat. What to do?

The race is on to address microplastics and other threats to marine life. Efforts to rid our waters of microplastic pollution range from engineered microorganisms that feast on the particles[420] to seaweed-based plastics that dissolve in salt water[421] to nano-scale magnetic coils that transform plastic into water and carbon dioxide.[422]

Will any of these substantially reduce the amount of plastic pollution? It's far too early to say. But every such effort gives reason for hope.

Significant efforts are underway to protect aquatic life, particularly in our oceans. In 2017 the UN General Assembly designated 2021-2030 as the UN Decade of Ocean Science for Sustainable Development (Ocean Decade). The goal, according to UNESCO, is to develop knowledge-based solutions to sustainably develop and maintain the marine ecosystems.[423] Halfway through the Ocean Decade, more than 800 governments, NGOs, universities, and scientists had pledged action on concerns as diverse as underwater noise pollution, reef restoration, and establishing marine protected areas.[424]

The international community has set a goal of protecting 30 percent of the world's waters by 2030. This "30X30" initiative, part of the Kunming-Montreal Global Biodiversity Framework, is a response to the alarming rate of marine species die off. It recognizes that immediate, coordinated effort is imperative if we are to reverse the trend and stave off ocean collapse.[425] Currently, marine protected areas encompass only 8 percent of the Earth's oceans, and most of that area is shielded in name only: A 2024 assessment found that a mere 2.7 percent had sufficient safeguards in place.[426]

Two notable initiatives aim to turn the tide. The first, the UN High Seas Treaty, seeks to protect international waters. These waters constitute some two thirds of our oceans. A hodgepodge of regional treaties protect some areas from some activities, but global regulations do not currently exist. As a result, the high seas have become a "maritime Wild West," as the UN Environment Programme puts it. Unregulated deep-sea mining, industrial fishing, and chemical pollution in international waters are among the threats pushing marine species toward extinction.[427]

The High Seas Treaty provides a framework for establishing marine protected areas, regulating commercial activity, and advancing knowledge through research and shared conservation strategies. More than 60 countries have now ratified the treaty, enough for it to become law.[428]

The second high-powered initiative takes aim at national waters. The presidents and prime ministers of 18 countries, including some of the world's largest maritime economies, formed the High Level Panel for a Sustainable Ocean Economy to drive "'bold transformations" that simultaneously protect ocean ecosystems and promote economic production and prosperity.[429] They have committed to sustainably develop 100 percent of their national waters. A sister organization, the Ocean Action 2030 coalition, provides technical and financial assistance to help Panel members develop and implement Sustainable Ocean Plans to guide economic and conservation activities.[430]

The Panel for a Sustainable Ocean Economy encourages all coastal and ocean nations to join its cause. After all, the boundaries between one nation's waters and another's are literally fluid. Sustaining the fisheries that provide sustenance for so much of the world's population—not to mention for polar bears, puffins, and seals—will require all nations uniting to protect our vital seas.

19

Agriculture

Thanks to Trader Joe's worldwide supply chain, I don't have to grow the ingredients of my favorite breakfast in my back forty. I've never done a nutritional analysis but, accompanied by a glass of vegetable juice, I find my breakfast of choice satisfying and filling. Best of all, it takes but a minute to prepare:

Lazy Lloyd's Granola

Combine in a small cereal bowl:

1 small handful of blueberries
3 soupspoons of raw oatmeal
1 small handful of walnuts
1 small handful of unsalted roasted peanuts
1 small handful of raisins
1 sliced banana

Add oat milk up to but not covering banana slices.

Confession: even if I could grow blueberries, oats, walnuts, peanuts, grapes, and bananas in my backyard, as you might have guessed, I'm too lazy. Thanks to farmers in North and Central America and millennia of agricultural innovation, it's not necessary.

And, yes, *mea culpa.* My breakfast comes at cost of greenhouse gas emissions; illustrates how even the most innocent pleasures contribute to our climate crisis. But, as things stand, there's no escaping it. Agriculture contributes between one-quarter and one-third of all global greenhouse gas emissions.[382]

Figure 19.1: Chickens

The story of agricultural innovation is the story of humanity's colonization of Earth.

Don't hold me to it, but by my reading our long lost relatives, *Homo erectus,* were using stone tools in the highlands of Ethiopia two million years ago[431] and, based on partial evidence, may well have been hosting barbecues in South Africa at least one million years ago.[432] They did get around, testament to their resilience and adaptability. They survived nearly two million years before fading out some 110,000 years ago.[433]

Why did *Homo erectus* fade away following such successful reign? University of Iowa paleoanthropologist Russell Ciochon and colleagues theorize that climate change did them in. They cite evidence that the last holdout of *Homo erectus* was in Java some 120,000 to 130,000 years ago. Much of Java had been open country habitat with abundant large mammals. But due to climate change, "...rainforest flora spread across Java. *Homo erectus* was not able to adapt."[434]

Homo erectus were hunter-gatherers.[435] Anthropologists, archaeologists, biologists, botanists, demographers, and linguists have studied the lives of contemporary hunter-gatherers for more than a century. They've found that hunter-gatherers possess vast knowledge of their environments. They know which plants of the many in their worlds are edible and when plants are ready to eat; which can be eaten raw and which need cooking. They know behaviors and life cycles of birds and animals. They make medicines, drugs, and intoxicants from native plants, as well as clothing, baskets, and cordage. They consume some 500 varieties of food harvested from their natural environment.[436]

While working toward his PhD degree in 1963, Richard Borshay Lee conducted field work among the !Kung San in northern Botswana. The !Kung are one of a dwindling number of hunter-gatherer cultures in the world. Lee reports that plant foods make up 60-80 percent of !Kung diet and, when available, meat is a special treat but never depended upon as a staple. The !Kung satisfy their need for food by working only two or three days per week. Working adults take responsibility for children, old people, and the disabled. Lee found that "...starvation, malnutrition, and crime were nil."[437]

Anthropologist Marshall Sahlins worked among indigenous people in Hawaii and Fiji. His observations agree with Lee's:

> ...hunters work less than we do, and rather than a grind the food quest is intermittent, leisure is abundant, and there is more sleep in the daytime per capita than in any other condition of society.[437]

This is not to advocate a return to foraging for edible plants and hunting wild game as the basis of subsistence. The hunter-gatherer lifestyle is only possible when population, human needs, climate, and the resiliency of nature are well balanced. This is no longer true for most regions of Earth, and hunter-gatherer survival skills are hard won and in short supply.

Today some 250,000 to 350,000 people subsist mainly from hunting and gathering across the Congo Basin,[438] 1,500 people live a foraging lifestyle in Tanzania,[439] and about 1,500 Lamalerans living on a remote Indonesian island support themselves by hunting sperm whales from small boats, using handmade harpoons.

Anthropologists note, however, that hunter-gatherers care about their environment. They do not hunt without need, they waste less, and they actively manage their resources. Hunter-gatherers living in different parts of the world, for instance, set controlled forest fires at fixed intervals to manage the landscape. Controlled fires help eliminate weeds and insects. They also promote the germination of seeds such as pine nuts, chestnuts, and walnuts that are encased in hard shells. This increases the number of seed-bearing plants that are available to eat.

The origins of agriculture are dimly perceived in the misty past. An international collaboration of researchers has found evidence that plant cultivation began some 23,000 years ago. But it's likely that humans relied on a mixed strategy of hunting, fishing, foraging, and some farming long before they took up cultivating the soil and herding goats full time.[440]

Since women did most of the foraging for food and medicinal plants, some believe that they invented agriculture. Women contributed hard physical labor, according to University of Cambridge researcher Alison Macintosh and colleagues. The researchers compared bone density of Central European women who lived during the first six millennia of farming with that of living female rowing champions and found that the farm women had stronger upper arms due to tilling soil, harvesting crops, and grinding grain to make flour.[441]

An odd thing happened to human reproduction some 4,000 to 8,000 years after agriculture became the predominant lifestyle:

> Across the globe, for every 17 women who were reproducing, passing on genes that are still around today—only one man did the same.[442]

University of Tartu (Estonia) researcher Monika Karmin and colleagues speculate that this resulted from cultural changes— rich farmers had more opportunities for sex than poor peasants and farmhands. But no worry, rate of reproductive sex between women and men eventually evened out.[443]

Bragging rights for first domestication of crops and animals are up for grabs. Archaeological evidence dating back 12,000 years turns up in diffuse areas around the world, including the Near East, China, Southeast Asia, Africa, Mesoamerica, South America, and northeastern North America.[444]

Cattle, sheep, and goats were first domesticated in the Near East 10,500 years ago.[445][446]

Pigs were domesticated in eastern Anatolia and central China 9,000-10,000 years ago.[447] Chickens were domesticated in South and Southeast Asia, southern China, Thailand, Burma, and India 8,000 years ago.[448]

Rice was cultivated in India some 6,500 years ago.[449]

Maize, beans, and squash were cultivated in South America and roots and tubers in Africa and East Asia 5,000 to 8,000 years ago.[450] Chocolate, the fruit of cacao beans, was consumed in Mesoamerica by Olmecs at least 3,500 ago.[451]

We can thank an observant Ethiopian shepherd named Kaldi for coffee. Legend goes that a century or so before Columbus set sail for the New World, Kaldi noted that his sheep were restless and refused to sleep after eating certain berries. He mentioned this to a priest at a local monastery. The priest made a drink from the berries that enabled him to stay awake through late-night prayer. The drink caught on with other monks and the rest is history.[452]

Early plows were made of wood and animal bones. Draft animals pulling plows enabled farmers to cultivate larger areas of land. Farm tools made of iron further advanced agricultural productivity; chemical fertilizers and mechanized tools powered by fossil fuels paved the way for sprawling corporate-owned farms.[453]

Hey, we could go on. The history of agriculture is rich and engrossing. Library shelves are filled with books, encyclopedias summarize high points, and archaeologists fill in details from every continent. Much of the story is about cultivation, amendment, depletion, neglect, and erosion of soil.

Soil is much more than dirt. It's a complex of minerals, dead and living organisms, air, and water. It's one of the Earth's most important natural resources.[454]

FAO tells us, however, that 90 percent of the Earth's topsoil is at risk— "...the equivalent of one soccer pitch of earth erodes every five seconds."[455] And it's not only erosion causing concern. Thousands of chemicals and tons of urban, industrial, and agricultural waste end up in soil, leaving it bare and degraded.[456] John Todd's book *Healing Earth: An Ecologist's Journey of Innovation and Environmental Stewardship* documents his institute's inspiring research and action to rehabilitate severely degraded land.[457]

Now is a good time to jump over to the Amazon rain forest for a look at *terra preta*, dark earth, which may well help address problems vexing both farmers and environmentalists today. Most Amazon rainforest soil is acidic and low in nutrients. Yet terra preta, fertile "dark earth," is found in patches across hundreds of sites. Researchers from the US and Brazil say the terra preta developed over centuries as ancient Amazonians enriched patches of the forest to nourish crops.[458]

Terra preta sites date back thousands of years and still cover some 10 percent of Amazonia. Anthropologists estimate that the productivity of terra preta cultivation could have supported populations as large as 100,000 people.[459]

MIT researchers believe that terra preta could be an effective recipe for sequestering carbon, as well as a technique for revitalizing unproductive land.[460]

One could say that the ancient Amazonians were environmental engineers well ahead of their time.

Today, half of the Earth's habitable land is devoted to agriculture. Of that, croplands make up one-third and grazing land the rest.[461] Yet some 33 percent of the world's soil is moderately to highly degraded.[462]

FAO's 2023 Statistical Yearbook World Food and Agriculture reports that:[463]

- Agriculture employed some 27 percent of the global workforce in 2021.
- Between 1991 and 2021 climate-related disasters caused production losses in crops and livestock valued at $3.8 trillion USD— some 5 percent of annual global agricultural GDP.
- Between 2000 and 2021, agricultural land declined by roughly twice the size of California.

The UN further reports that:

> Almost 10 percent of the 8 billion people on Earth are already undernourished, with 3 billion lacking healthy diets, and the land and water resources farmers rely on are stressed to 'a breaking point.' And by 2050 there will be 2 billion more mouths to feed.[464]

The Global Food Banking Network reports that members "provided 1.7 billion meals in 2022 to more than 40 million people struggling with hunger, reaching near historic levels of assistance not seen since the height of the COVID pandemic."[465]

Farmers around the world are hard pressed to feed a growing population.

Globally, some 74 percent of land livestock are factory-farmed.[466] In polite company, factory farms are called "concentrated animal feeding operations." On the ground, they crowd pigs, cows, sheep, turkeys, chickens, and ducks into pens and cages for months to a year at a time, with significant stress on the animals. Their goal is to maximize profit with minimal expenditure of resources.[467]

The stress of overcrowding suppresses the animals' immune response, and the close quarters increase parasite loads. Compromised immune systems make animals more susceptible to infection and contribute to mutation of disease-causing bacteria such as *Salmonella* and *E. coli* and viruses such as those that cause avian and swine flu. Some of these diseases can be transmitted to humans. Long-term overuse of antibiotics to enhance growth and prevent disease in animals contributes to the rise of antibiotic-resistant bacteria.[468]

European researchers conducted an in-depth literature review of the impact of intensive poultry farming on the environment and human health. They found that poultry litter and manure can lead to air, soil, and water pollution with adverse effects. Their review concludes with measures industrial farms, feed and pharmaceutical producers, and regulatory authorities will need to take to address the challenges of public health, population growth, climate change, and resource conservation.[469]

Nevertheless, the US Department of Agriculture reports that the number of large factory farms is on the rise. The decade from 2012 to 2022 saw an increase of mega-factory hog farms from 3,006 to 3,540 and a 29 percent increase of cattle farms with more than 5,000 cattle. Chicken farms with 500,000 or more birds increased by 17 percent.[467]

The number of animals raised in small spaces saw similar growth over the decade— 28 percent more cattle, 24 percent more hogs, and nearly 1.4 billion more chickens.[470]

The vast majority of factory farms in the US are owned by small farmers operating under contract to large corporations such as Tyson, Pilgrim's Pride, and Perdue. Take chicken farming, for instance. Chicken farmers who contract with large corporations raise some 95.5 percent of all chickens consumed in the US. To enter into such contracts the chicken factory farmers must build facilities to exact company specifications. Many take out loans ranging up to a million dollars. They're frequently asked to make costly upgrades requiring additional loans. Factory farm work is physically hard. Yet, according to the Department of Agriculture, median net revenue for poultry farmers in 2022 was $9,367.[471] According to poultry farmer Craig Watts, "The debt keeps them trapped in the system."[472]

Tyler Whitley, director of the Transfarmation Project, calls factory farming "a cancer on rural America." Whitney's organization is working to help factory farmers transform their farms into more profitable and less stressful operations. "I've never heard of a factory farmer who's regretted their choice to get out," Mr. Whitley said. "The longer that they're out, the happier they are."[473]

As I write, 2025 has been particularly hard for farmers overall. They've endured lower commodity prices; higher costs of seed, fertilizer, pest management tools, and diesel; trade disruptions; and unfavorable weather. The first three months saw 259 farm bankruptcies— more than the entire year of 2024.[474]

So what are the options?

Innovative farmers are moving toward science-based practices that maximize productivity and profit while minimizing environmental damage. Called regenerative farming, the movement seeks social, economic, and environmental sustainability. Adapted to local conditions and regional markets, regenerative farming supports farms of all sizes and a diverse range of foods, fibers, and fuels.[475] In short, sustainable farming practices seek to promote soil health and minimize water use and pollution levels while enabling farm workers to make a better living.

But the big question is this:

How can we sustainably feed 10 billion people by 2050?

The World Resources Institute has published 22 recommendations that cut across the entire food chain and more, suggesting that we're up against an all-hands-on-deck challenge:[476]

Policy

- Adapt to climate change
- Link agricultural productivity with protection of natural ecosystems
- Avoid competition between food crops and bioenergy for land
- Improve wild fisheries management

Research

- Improve crop breeding
- Increase livestock and pasture productivity
- Improve productivity and environmental performance of aquaculture
- Reduce enteric fermentation (digestive process of ruminant animals) through new technologies

Farming

- Increase agricultural energy efficiency and shift to non-fossil energy sources
- Improve soil and water management
- Implement realistic options to sequester carbon in soils
- Plant existing cropland more frequently
- Reforest agricultural lands with little intensification potential
- Reduce emissions through improved manure management
- Reduce emissions from manure left on pasture
- Reduce emissions from fertilizers by increasing nitrogen use efficiency
- Limit inevitable cropland expansion to lands with low environmental opportunity costs

- Conserve and restore peatlands
- Adopt emissions-reducing rice management and varieties

Supply Chain

- Reduce food loss and waste

Consumption

- Shift to healthier, more sustainable diets

Population

- Achieve replacement-level fertility rates— on average 2.1 children or fewer per family

No doubt there's even more to be done across the board, and success rests upon details of implementation.

The Global Hunger Index, published by Concern Worldwide, Welthungerhilfe, and the Institute for International Law of Peace and Armed Conflict, defines low hunger as ≤ 9 on a 100-point scale. The lower on the index the better. As of 2024 the GHI is 18.3. "If progress remains at the pace observed since 2016," they state, "low hunger at global level may not be reached until 2160."[477]

Good news: The Union of Concerned Scientists reports evidence that farming practices that emphasize building healthy soil, preventing erosion, managing water wisely, and other sustainability measures can be just as productive and profitable as traditional agricultural practices and, over time, meet global food needs.[478]

Irish farmer and sculptor Eoghan Daltun argues that healthy natural ecosystems are an essential foundation for sustainable agriculture. Climate and ecological breakdown are a major threat to the future of farming, he says. Stable climate makes farming possible and a stable climate depends upon natural ecosystems.[479]

So Daltun has taken matters into his own hands by rewilding 73 acres on Ireland's Beara Peninsula, turning over-grazed land back to Atlantic rainforest. Forests once covered 80 percent of Ireland. Today they cover just 11 percent and, of that, native trees cover just one percent. Daltun concludes:[479]

> It's not just by burning fossil fuels that we're destroying the stability of the climate. It's also by erasing natural ecosystems

which regulate the climate. We need to start looking at the big picture here, which is that everything depends on nature.

20

Mining

The United States Geological Survey reports that 54 minerals are essential to US energy, defense, and technology industries.[480] The list ranges from aluminum to zirconium.

Aluminum is the second most abundant metallic element in the Earth's crust.[481] Aluminum has many consumer and industrial applications where low density, thermal conductivity, and corrosion resistance are prized.

Zirconium has invaluable properties for industrial, medical, and consumer applications. The most valuable source of zirconium is zircon ($ZrSiO_4$), which is found in Australia, South Africa and India.[482]

IEA cites the relative importance of nine critical minerals across 10 clean energy technologies:[483]

- Copper
- Cobalt
- Nickel
- Lithium
- Rare Earth Elements
- Chromium
- Zinc
- Platinum Group Elements

Figure 20.1: Chuquicamata — one of the world's largest copper mines
Credit: Owen Cliffe, CC BY-SA 3.0

- Aluminum

To meet global climate neutrality by 2050, the demand for these minerals is expected to expand by as much as six times, according to the IEA.[484] Extracting and processing critical metals, however, raise challenging environmental, geopolitical, and sustainability issues.

Mining generates millions of gallons of wastewater, a major source of water pollution. Water within and around mines is usually contaminated. Treating mine wastewater before it is released into the environment is expensive and all too often neglected. Indeed, most of the material coming out of mines is regarded by miners as waste— earth and stone removed to reach the ore and tailings left over after valuable ore is separated out.[485] Tidbit: it requires up to 12 tons of ore to extract one troy ounce of pure platinum.[486]

Geopolitical issues popping up around mining include trade disputes, supply chain disruption, and human rights abuses. Consider:

— China dominates processing of global clean energy minerals.[487] In 2023 China's Ministry of Commerce and General Administration of Customs imposed restrictions on the export of gallium and germanium to "protect national security."[488] China threatened to restrict even more critical metals if trade issues were not resolved. Western nations have mobilized financial resources to break China's chokehold over critical metals much needed by high-tech industries. But the dispute is yet to be resolved.[489]

— The Democratic Republic of the Congo extracts 70 percent of the world's cobalt supply. Slavery, child labor, and families forcibly displaced from illegal mine sites are among the many human rights abuses that abound across the DRC cobalt mining industry.[490]

— Indonesia produced an estimated 40.2 percent of the world's nickel in 2023.[491] China has significant investments in Indonesian nickel reserves and processing industry. But Indonesia is seeking access to nickel markets in the US. The US Department of Labor includes Indonesian nickel in a list of products that employ forced labor.[492]

Minerals are nonrenewable elements. Once extracted from the Earth there is no more to be found until mining operations on the moon or elsewhere in the solar system become economically feasible.[493] Science fiction fantasy? Houston, Texas, start-up Lunar Resources plans to build the largest telescope in the solar system on the far side of the moon. They see securing material on-site to build the telescope project as R&D and demonstration of off-planet mining and manufacturing.[494]

But in the shorter term, here on Earth, seawater is a promising source of minerals and metals. Sodium chloride, bromine, and magnesium salts are currently extracted from desalination concentrate.[495] With advanced technologies it may be possible to extract many other minerals, including uranium.

Tidal Metals, a startup company founded by three PhD scientists from MIT, Princeton University, and the University of Wisconsin-Madison, is developing an evaporative technology for extracting magnesium from desalination brine. Their process recycles 97 percent of the energy involved while evaporating one metric ton of seawater to produce 4 kilograms of the magnesium salt. "The one desalination plant in San Diego pumps in 100 million gallons of seawater a day. That's enough to supply all of the US with magnesium— 180,000 tons a year," says Howard Yuh, co-founder and CEO of Tidal Metals. "Today, it's all going to waste. The name of the game really is, can you compete with the 90% production that's coming out of China today?"[496]

Down on the farm, well, a rented plot of land in a rural village on the Malaysian side of the island of Borneo, University of Melbourne visiting botany professor Alan Baker and an international team of colleagues are demonstrating how nickel can be "farmed" on toxic soils and mine tailings. Turns out that some 700 species of plants flourish in mineral-rich soils. Many take up and "hyperaccumulate" metals through their roots.

Professor Baker comments that a farmer can earn thousands of dollars every six-to-12 months by selling nickel citrate on international markets. To produce some 500 pounds of nickel citrate, the farmer would shave one foot of growth off of a crop of nickel-accumulating plants and run the harvest through a relatively simple extraction and purification process.[497]

Extracting minerals from hyperaccumulating plants is called phytomining. It's unlikely to replace traditional mining, but it may help contribute to mineral supplies, clean up toxic soils, and provide supplemental income to small farmers.

Researchers at the University of Edinburgh are using microbes to recycle lithium, cobalt, and other expensive minerals from old batteries and discarded electronic equipment.[498]

Deep-sea mining is another possible source of critical metals, including:[499]

- Potato-size manganese nodules on the ocean floor (rich in manganese, cobalt, copper, nickel, and rare earth elements)
- Cobalt-rich crusts lining the sides of mid-ocean ridges and underwater mountains, also known as seamounts

Environmentalists argue, however, that the impact of deep-sea mining is not well-enough understood to justify mining operations.[500] Legal experts agree that a moratorium on seabed mining is legally required by the UN Law of the Sea.[501]

Long-term sustainability of critical metals, however, will depend on efficient recycling in a circular economy[502] and discovery of functional substitutions.[503] Addressing social justice issues in mining will require global awareness and political solutions.

The UN Secretary-General's Panel on Critical Energy Transition Minerals is working to "build trust between governments, local communities and industry, by addressing issues relating to equity, transparency, investment, sustainability and human rights."[504]

The Panel's report, released on September 11, 2024, outlines measures to embed equity and justice in the race to net-zero emissions.

21

Circles

Walking through a deciduous forest on a beautiful fall day we're struck by the waste. Leaves! Leaves discarded everywhere! Ankle deep on the ground. One would think that discarded leaves would bury the forest in a matter of years.

But we know that won't happen. Nature has evolved leaves to be upcycled into plant nutrients and soil enrichments which, in turn, nourish trees and other living things in the forest.

Round and round, season upon season, nitrogen, phosphorus, potassium, sugars, amino acids, and other essential plant nutrients cycle through the soil, forest plants, trees and leaves, and back to the soil.

The same principle applied to material production, consumption, waste, and recycling is called a circular economy.

Optimistic visions of circular economies are drawing considerable attention across governments and industry. The idea has been circulating since 1976, if not earlier.[505] But nature evolved circular economies long before humans gave them a thought.

The Ellen MacArthur Foundation defines a circular economy as "...a systemic approach to economic development designed to benefit businesses, society, and the environment."[506]

Goals:

- Eliminate waste and pollution
- Circulate products and materials

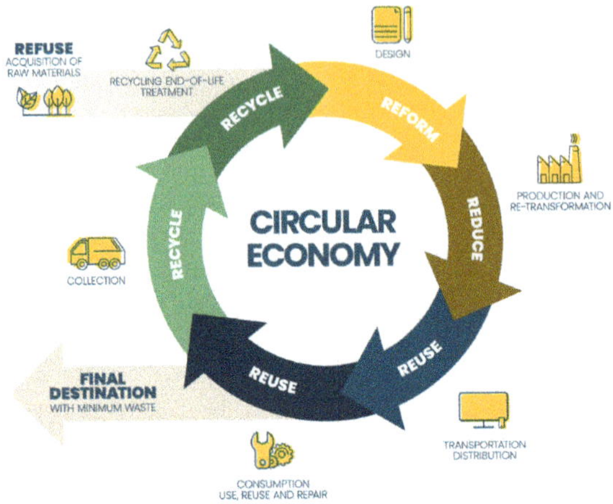

Figure 21.1: Circular Economy
Credit: iStock

- Regenerate nature

The World Resource Institute tells us that more than 100 billion tons of metals, minerals, fossil fuels, and organic materials enter our global economy every year:

> Just 8.6% gets recycled and used again. Use of resources has tripled since 1970 and could double again by 2050 if business continues as usual. We would need 1.5 Earths to sustainably support our current resource use.[507]

Transition from our linear extract-consume-discard economies to circular economies would deliver significant benefits:

- Make better use of finite resources. Take textiles for instance. Production of synthetic fibers consumes huge quantities of fossil fuels. Production of natural textiles, such as cotton, consumes some 100 billion cubic feet of water, roughly 4 percent of global freshwater supply. People throw away still-wearable clothes worth an estimated $460 billion USD each year. Purchase of 100 second-hand garments could cut production of 85 new ones.[507] Circular economies reduce use of resources; could cut global greenhouse gas emissions by as much as 39 percent (22.8 billion tons).[508]

- Reduce emissions. Nearly 45 percent of global greenhouse emissions come from product use, manufacturing, and food production.
- Protect human health and biodiversity. Air, water, and soil pollution contribute to more than 9 million deaths globally each year and to decline of biodiversity. Circular economies help protect human health and biodiversity by making better use of natural resources (e.g., protecting water and land) and by mitigating the climate crisis.
- Boost economies through waste reduction and job creation. Moving toward a circular economy is a $4.5 trillion economic opportunity.
- Create more and better jobs in product design, manufacturing, repair services, remanufacturing, and construction.

Despite growing enthusiasm for circular economies, however, the share of global materials used, recycled, and sold for reuse has decreased from 9.1 percent in 2018 to 7.2 percent in 2023 — a 21 percent drop.[509] That said, the circular economy is growing. The challenge is that humanity consumes more materials than recyclers can process. Over the six-year period from 2018 through 2023 we consumed 28 percent of all the materials that humanity has consumed since 1900. What to do?

What can we do?

Share, buy used, repair, and recycle.

Can it be that we've come full circle?

Early in this book we posed two questions:

1. What kind of world do we want for our children and theirs?
2. How must we manage affairs on spaceship Earth to achieve it?

Nature recycles energy and spent materials through complex, intertwined networks. Plants and animals in the network convey, store, and release energy to combine elements such as nitrogen, oxygen, carbon, and sulfur into molecules such as carbohydrates and proteins and break them down again— what Kate Raworth calls a "network of flows." Nature has evolved sustainably for more than 4.5 billion years. Humanity is one of countless manifestations of nature's creative fecundity and power. But somehow humanity has slipped out of sync;

disrupted the delicate balance that stabilizes a climate on Earth favorable to living things— including ourselves, our children, and theirs.

Meanwhile, we've over-exploited natural resources and hindered nature's capacity to regenerate life support systems that we take for granted— breathable air, fresh water, nutritious food, and materials at the base of our economies.

Our challenge, then, is to learn from nature, to transform our lifestyles and economies from linear exploit-consume-discard to circular harvest-use-repair-reuse-recycle-regenerate, and to cultivate economies that give back to nature as much or more than we exploit.

Kate Raworth's doughnut model points the way.

It's up to us, all of humanity, to navigate Spaceship Earth back on course toward a sustainable future for our children and theirs.

22

What to Do?

December 22, 2024

FLASH!

Figure 22.1: Palisades Fire 2025
Credit: Jessica Christian, Unsplash

Fire is devastating Los Angeles County: "...nuclear-scale destruction...block after block and neighborhood after neighborhood leveled — roughly 12,000 structures destroyed or rendered uninhabitable, 55 sq. miles of city and mountains burnt, nearly 200,000 residents evacuated — so far."[510]

Neighborhoods, affluent and needful— incinerated. Homes, where people once loved and children played, mansions to mobile— ashes. People, long inspired by hopes and dreams, rich and poor— fleeing before advancing flames.

Is this the new normal?

What did these people do to deserve this? Have I, in careless ways large or small, contributed?

I can't write. Too distressing.

January 12, 2025

LA wildfires continue to burn. More high winds forecast. Power off in evacuated areas. Kenneth fire in the west hills is 100 percent contained.

January 13, 2025

The Palisades fire is 14 percent contained. The Eaton fire is 33 percent contained. The Hurst fire, 97 percent contained. More than 15,000 firefighters are working around the clock to bring multiple fires under control.

AccuWeather estimates that damages and economic losses of the 2025 Los Angeles fires will exceed $250 billion USD.[511]

The Los Angeles fires mark just one more tragedy in a decades-long trend of ever more destructive weather events. The World Meteorological Organization reports that some 12,000 extreme weather disasters have caused two million deaths and $4.3 trillion USD in economic losses over the past half-century.[512]

Post Carbon Institute Senior Fellow Richard Heinberg tells us that we face a choice between a gentle path forward or catastrophe and suffering.[513]

Did the Angelenos who've lost their homes, schools, churches, and local places of business have a choice? Or the people who died?

The gentle path, says Heinberg, requires systematically giving up "...many material features of modernity and bringing down fertility rates cooperatively and peacefully...while retaining whatever serves life, beauty, and happiness."

Giving up?

Is Richard Heinberg advising us to forego comfort, convenience, and security? Wealth and standing? This is a tall order and a big ask of humanity. Humanity. That's you, me, and more than eight billion others.

I'm more optimistic than Heinberg. But I can't argue that he's wrong. The Los Angeles fires, Hurricane Helene, the 2025 Texas floods, and scores of climate-implicated catastrophes around the world bear testimony. Truth is, nature won't wait for us to resolve differences, make up our minds, and settle on a path— gentle or not.

And this gentle path he proposes, what does it mean? Who is he asking to give up what? Me? You? Rich people? The poor? Humanity?

We circle back to agency— the power to act.

We each make choices, countless, from the moment of wakefulness 'til our minds shutter in sleep. What to wear, what to eat, where to go, what to buy, who to seek out, who to shun, who to believe, what to believe, how to invest, in what, and how. But our choices are constrained by family, financial means, community values and norms, and law. Our mobility is constrained by walls, paths, roadways, and means of transport. Our actions are facilitated or impeded by others. As some have said, "Your freedom ends where my nose begins."

And more, we're embedded in economic, technical, political, and cultural systems. Our ability to navigate and exploit complex codes — finance, computer systems, law, social manners and customs — goes far toward determining our success in life.

As human beings we're social animals. We're imitative. Conformist. Doing what we need to survive within status and power hierarchies. And now, environmentalists and nature itself are telling us that we must pivot on a dime.

Changing our personal beliefs, spending habits, and behavior is challenging enough. How many New Year's resolutions are abandoned and forgotten by the ides of March? Reforging economic and political systems at global scale may well feel like a pipe dream.

So what to do?

That's the question, isn't it? So many questions...

— Who cares and why?

— What kind of world do I want to live in?

— How do I want to live my life?

— How do I want my children and grandchildren to remember me?

— What kind of world do I hope they will inherit?

— What kind of lives do I hope they'll have?

— Do I care?

We each have our own struggles, resources, talents, dispositions, and opportunities. We can only do what we can do. But, as Socrates said, "To know thyself is the beginning of wisdom." So, as individuals, we can introspect:

— What changes in my life would bring a greater sense of fulfillment and meaning?

— How can I buy, consume, and accumulate fewer material goods while enjoying life more?

— How can I help my children and grandchildren understand and respect the natural world and life itself?

— What do I truly value in the people, institutions, and natural environments that surround and shape my day-to-day?

— How can I support people and organizations working toward a more just and habitable world?

— How can I be a better steward of the natural world around me?

— How can I improve my physical health and well-being to be the best that I can be?

Insight through introspection is hard won. The paths forward are countless. But which to choose? Once we see the way, however, we can learn and do. It helps to know that we're not alone. The gentle path is a journey of multitudes joining hands in mutual cooperation and respect. Many have forged the path before us. And we, in turn, have power to inspire others who join the arduous path toward justice and sustainability in our wake. We've all heard it, "The journey of a thousand miles starts with a single step."

Starting at home and in our communities we can:

— Uphold our core environmental values through behavior and action.

— Inspire our families and all we know with our love and respect for the natural world.

— Work with others to clear the path toward a more just, habitable, lush, and beautiful future.

There's nothing new here. Prophets, seers, philosophers, and the wisest of the wise have helped us bond in civility for millennia. But philosophers of yore could never have imagined how significantly the globe-spanning Internet and social media have transformed our lives and social relations.

The Internet can be a powerful resource to help us along the gentle path or— it can distract us, lead us perilously astray.

Have you ever noticed how many people are scrolling through their mobile phones in restaurants or other public places, quite oblivious to others around them? Tech company founder Mason Yarbrough advises aspiring entrepreneurs to "...build systems that make social media less necessary. Improve third spaces where people connect directly, authentically, without intermediation by systems designed to extract value from their attention."[514]

The gentle path might teach us that what was once old is new again.

In the here-and-now of our mediated world, however, we can:

— Reduce screen time on social media and spend more face time with family and friends.

— Critically evaluate all information we find on the web. Check credibility of sources, fact-check questionable information, and cross-check with multiple sources.

— Post information and ideas through social media, blog posts, and websites to help others navigate their way along the gentle path.

This brings us to the big show.

Addressing climate and habitability issues at global scale is indeed a formidable challenge. But my take is that we have to fight for a just, habitable, indeed, survivable, world with every fiber of our being. Bottom line: There's little time to pivot away from the path of catastrophe and suffering; too much to lose should we fail.

The good news is that the gentle path is a numbers game. If enough of us stand up, learn, educate, advocate, demand, live, and work toward a more just and supportive future, we can significantly up the odds toward long-term habitability. Many smart people around the world have dedicated their lives toward this goal. When we stand up for the future of humanity, we aren't alone. We're in good company

All of us, however, must keep laser-focused on our own personal values, choices, and actions; how they fit into the big picture of climate change and long-term sustainability of humanity.

As the great primatologist Jane Goodall reminded us, "You cannot get through a single day without having an impact on the world around you. What you do makes a difference, and you have to decide what kind of difference you want to make."[515]

While mounting offense against forces propelling us toward catastrophe and suffering, however, we must not neglect defense. Ever more extreme weather is inescapable. We must innovate and adapt to extreme weather already baked into our climate and mitigate damage as we go.

Yes, ignorance, greed, envy, fanaticism, and entrenched power stand in our way.

Sociopaths and psychopaths rise to power, wage wars, exploit and oppress the weak. We must understand how they come to be, recognize them early in their development, and keep their hands off consequential levers of power. It's a hard go. A fight. That's where psychology, sociology, political action, and voting come in.

Let's now consider top priorities.

With utmost urgency we must reverse global heating. If we don't win the battle to suppress greenhouse gases, we lose the war for the sustainability of human life on Earth. This can't be done within a week, month, or year. Some 32 million employees rely on revenue from fossil fuel production, distribution, and sales to feed their families.[516] Some 7,500 institutional investors hold $5.1 trillion USD in fossil fuel shares and bonds.[517] Even with ample sincerity and will, transitioning from oil, gas, and coal to 100 percent renewable energy no doubt feels to fossil fuel executives as impossible as copping a U-Turn from the bridge of the 380-meter OCEANIA oil tanker midway through the Suez Canal.

But imagine that the massive OCEANIA is fast closing in on a deftly hidden nuclear mine. Given the length and width of the ship and the narrow constraint of the waterway, it's clearly impossible to turn the ship around.

What would we have the captain do?

The captain would certainly want to protect his own life and has a duty to protect the lives of his crew.

The CEO of the company that owns the ship would want to protect the valuable ship and monetize the value of the oil onboard.

Investors in the shipping company have a fiduciary obligation to maximize the portfolio value of their investments.

What to do?

Options:

1. Deny the report of a nuclear mine as fake news
2. Let the ship blow and collect the insurance
3. Bring the ship to full stop and hire a marine towing company to tow the ship back to open water
4. Recruit nuclear dismantlement specialists to disarm the mine— if they can find it

The choice rests on all-too-human beliefs and values. Blind faith. Willful ignorance. Narcissistic greed. Respect for human lives and long-term consequences. Stakeholders must choose one and act— or not.

All-too-many influential corporate and political leaders seem to value short-term gain over mitigating the long-term costs and consequences of global heating. Yet prudent leadership is essential to minimize future costs and tragic consequences.

At minimum the transition from fossil fuels to renewables requires:

1. Vision
2. Alternative technologies
3. Funding for alternative energy investments
4. Leadership
5. Planning
6. Global cooperation

Renewable energy technologies are at hand and advancing. Within the decade leading up to 2024 the cost of photovoltaics dropped by 90 percent, wind energy 70 percent, and batteries more than 90 percent. And market share of renewable energy sources is on the rise. Indeed, in 2023 renewables provided 30 percent of the world's electricity.[518] IEA reports that almost one-in-five cars sold in 2023 was electric.[519]

And it's fair to say that financial resources abound.

UN Trade and Development estimates that it will cost more than \$11 trillion USD annually to meet the UN Environmental and Energy Sustainable Development goals— \$5.5 trillion USD invested annually to fight climate change, protect biodiversity, and cut pollution; \$5.8 trillion USD per year toward transition to clean energy.[520]

But consider that, globally, governments subsidized fossil fuel corporations and projects by the equivalent of \$7 trillion USD in a single year.[137] Subsidies include both payment out of public funds (\$1.3 trillion USD) and environmental costs imposed on the public. Multi-billion to trillion dollar subsidies are not a one-off. They're bequeathed annually at the expense of life on Earth. Qatar provides the highest subsidy— more than \$2,500 USD equivalent per capita. IMF projects that global fossil fuel subsidies could be as much as \$8.2 trillion in 2030 due to rising fossil fuel consumption.

The cost of infrastructure to transition to clean energy is a one-time investment. Once infrastructure is built out, fuel costs largely disappear and savings accumulate.

Given liability for harms inflicted on individuals, communities, ecosystems, and future generations, repurposing fossil fuel subsidies to cover transition costs should be a no-brainer. At minimum, fossil fuel resources invested in exploration, research, engineering, and logistics should be repurposed to accelerate the transition to clean energy.

Advocate and agitate to make it happen.

As to transition frameworks and plans— the UN,[521] PennState Institute of Energy and the Environment,[522] Rocky Mountain Institute,[523] and others have outlined strategies and steps to move from fossil fuel-based energy systems to renewables. The devil is in the details. Actionable plans must address the interests of fossil fuel industry stakeholders— everyone from pension-fund investors to refinery workers to corner service station attendants and mechanics.

Is it possible?

Maybe. It may require powerful incentives. Imagine replacement of subsidies with phased buy-out of coal, oil, and gas reserves combined with consumption taxes.

Every loss on the fossil fuel side of the ledger, however, including job loss, must be paired or exceeded with gain on the renewables side. This won't happen without government/business cooperation and a big swing away from fossil fuels by consumers. And the transition from fossil fuels to clean energy must be global.

Note that as of 2025 the total market cap of the world's 100 largest oil and gas companies was just shy of $7 trillion USD.[524] Yet, as noted, subsidies and economic losses worldwide for oil, coal, and natural gas hit $7 trillion USD. What's wrong with this picture? If the numbers hold and I've got it right, nation-states could simply buy the fossil fuel companies outright and redirect research, engineering capacity, and profits toward clean energy and extreme-weather mitigation.

Fanciful?

It would take a small team of economists to project the cost/benefits of fossil fuel industry buy-outs. Among other things, they'd need to consider alternative scenarios for achieving 100 percent clean energy. National Renewable Energy Laboratory researchers have already looked at supply-side options. They found that benefits of investing in clean energy are greater than costs.[525]

As noted earlier in this chapter, the UN and other agencies have estimated the cost of 100 percent transition to clean energy. With advances in technology, actual costs of transition may well decline. The remaining questions center on the capacity of fossil fuel industry research and engineering resources to accelerate the transition. With existing reserves, the industry would continue to generate revenue, although profits would likely decline in step with decreased demand for oil and gas. If nothing else, global buyout of fossil fuel companies and repurposing research and engineering resources may be a way to soften the blow of job loss.

Too ambitious?

Consider the transition of the US economy to wartime production in 1942 and back to peacetime in 1945. Even more notable— consider the many countries that rebuilt whole cloth from rubble to prosperity. With financial aid, technical help, and infrastructure support provided by governments and the private sector, Western European economies shattered by the war returned to pre-war levels of output in less than five years.[526] The US's Marshall Plan contributed $13.3 billion in aid to 16 countries between 1948 and 1951, roughly $150 billion in today's dollars.[527] US assistance to Japan over that period was roughly $15.2 billion, $253 billion in 2025 dollars.

So it comes down to intent and political will. We can thank the fossil fuel industry for giving the world more than a century of cheap energy that has powered our industries and economies, inspired invention, and enriched our lives. But we know now the costs. And, to our good fortune, should we grasp it, we have the know-how, technologies, and resources to expeditiously transition to renewables.

Yes, fossil fuel decision makers and their political hired guns stand squarely in the way. The fossil fuel industry spends millions to influence public opinion and to buy political protection—$201 million USD annually by the top five oil companies to lobby the US Congress; some $750 million USD each year to portray themselves as proactively concerned about climate change.[528]

We can't allow the entrenched power of the industry to imperil future generations. Consider: We're the consumers. Energy companies exist to serve us. We have power to express our preference for renewable energy sources through day-to-day choices. If fossil fuels are the only choice, then we must conserve—consume less. But renewables are ever less expensive and ever more abundant and available. We must make our preference known with utmost determination through every available channel— personal, cultural, political, and legal.

In short, as energy consumers we bear collective responsibility for the future of humanity on Earth. It's our voracious appetite for more, more, faster and faster that's driving atmospheric instability, species extinctions, and collapse of ecosystems. We must curb our appetites; conserve, preserve, restore, innovate, transform, and recycle. We must respect the resiliency of nature— live within restorative constraints and reject violent conflict as the mediator of geopolitical differences.

Phase down of fossil-fuel consumption, however, is not enough to keep us on the gentle path. Of equal urgency is protection of fertile soil and rehabilitation of degraded farmland worldwide. Farmers and the food industry must take the lead here with sustainable farming practices[529] and innovation to minimize food waste throughout the supply chain. We have a part to play as well through our food choices, preparation, leftover management, activism, and discretionary buying power.

While writing "we," I'm referring to the most voracious consumers among us. We know that far too many people around the world have barely enough food and shelter to survive, while a few individuals at the top of the economic pecking order have more income and wealth than many nation-states.[530] Then there's the rest of us, fortunate enough never to worry about our next meal, careless enough to forget leftovers pushed to the back of the fridge. Can we make better choices? Move toward greater equity?

Even with optimal farming practices and equitable distribution of agricultural produce, can our finite Earth provide enough food to go around? Clearly that depends upon how many mouths we need to feed.

This swings us back to Richard Heinberg's second precondition for negotiating the gentle path— bringing down fertility rates. The debate over human numbers on Earth is long in the tooth, still contentious, but vital.[531]

If global TFR drops below 2.1, population on Earth begins to decline. If it rises above, population grows. In 2024 epidemiological researchers reported that TFR had dropped globally from 3.61 in 1980 to 2.23 in 2021. They projected that it would drop below 2.1 by 2030.[532]

TFR is not consistent across nations. As of 2024, Taiwan's TFR was 1.11, whereas Niger experienced a TFR of 6.64.[533] High TFR means more mouths to feed, heavier demands on women, and greater demand on natural resources. Lower TFR means fewer young people to support the elderly.[534]

What to do?

That's for women and their partners to decide. We must keep laser-focused on our own personal values, choices, and actions; how they fit into the big picture of climate change and long-term sustainability of humanity.

So what can we do?

Consider Greta Thunberg.[535]

What did she know?

— Climate change was a threat to all she cared about.

When did she know it?

— Age 15, if not sooner.

What did she do about it?

—Skipped school on Fridays to sit down in protest outside the Swedish Parliament.

How did she know what to do?

— She knew little more than any three-year-old— make a fuss and say, "NO!"

What did she accomplish?

— Focused worldwide attention on the imminent threat of climate change.

— Inspired a global student strike movement called Fridays for Future.[536]

— Stood before international forums and challenged world leaders to take immediate action to reduce greenhouse gas emissions.

Do student movements make a difference?

— As my grandmother-in-law would say about chicken soup as a cure for the common cold, "Can't hurt!"

Today, youth groups in more than 184 countries are striving to build a sustainable future through advocacy and action. More than 100 thousand young people around the world collaborate across nation-state boundaries through the UNESCO Youth Climate Action Network.[537]

Consider now the 19 students from the Pacific Islands who reshaped global climate law.[538]

What did they know?

— They'd seen beaches washed away by the rising sea. They'd seen vibrant coral reefs lose brilliant colors and turn white. They'd experienced decline of fish populations that their families and communities depended upon for protein. They'd lived through ever more frequent tropical cyclones that destroyed their homes and schools.

When did they know it?

— They knew the devastation of climate change well before they logged into their University of the South Pacific online law class in 2019.

What did they do about it?

— They studied the science; asked, who's responsible for the rising seas eroding our beaches, the coral bleaching, the loss of fish, and the ever more frequent cyclones destroying our homes and schools?

— They created an organization called Pacific Island Students Fighting Climate Change and coined a slogan: "We are taking the world's biggest problem to the world's highest court!"

— Inspired by a class assignment, they wrote a letter and a legal brief advocating for intervention by the International Court of Justice (ICJ).

— They presented their case to island leaders at an official meeting of the Pacific Islands Forum. Their ask: "...endorse at the highest level our fundamental moral and legal rights to live with dignity without fear of climate change denying us and our children of that freedom." Their proposal was noted by the Forum but not endorsed.

— They reached out to legal experts. A fool's errand, some said. But they persevered.

— Their organization, Pacific Island Students Fighting Climate Change, allied with another— World's Youth for Climate Justice. They were now at the nucleus of a movement. The movement inspired leaders of countries far beyond the Pacific, attracting support from Caribbean and African nations.

— A core group of 18 countries from regions around the world collaborated to push a resolution through the UN General Assembly requesting an IJC advisory opinion on the obligations of nations with respect to climate change. The resolution passed. One lawyer called the oral arguments before the IJC "...the largest legal proceedings ever in human history."

What did they accomplish?

— On July 23, 2025, the court ruled that the full range of international law applies to climate change. The decision was unanimous— climate commitments made during international negotiations are binding. Countries have a duty to prevent environmental harm and to hold those who cause harm legally accountable.

How did they know what to do?

— They didn't. But inspired by a clear and urgent goal, and with help from their law professors and other knowledgeable and influential people, they persevered and muddled through.

Did they make a difference?

Most likely. The ICJ decision supported their concerns and rewarded their advocacy. The legitimacy of law is based on legal precedence. The unanimous ICJ decision established a powerful legal precedent.

Many say that we are at a legal turning point. For instance:

In 2024 the International Tribunal for the Law of the Sea issued an advisory opinion that states have a legal duty to prevent greenhouse-gas pollution from harming the oceans. The opinion echos what the University of the South Pacific students were asking for in 2019.[539]

This year, 2025, the Inter-American Court of Human Rights issued an opinion that countries must protect the climate system on which human rights depend.[540]

A petition for a similar opinion is currently before the African Court on Human and Peoples' Rights.[541]

The paths forward are many. So I'd say, roll up our sleeves. Follow our hearts. Check out Appendix C to contact organizations worthy of support. Embrace creativity, adaptability, and our higher nature. Join the many around the globe who've already chosen the gentle path toward a habitable future. And more, reconsider every lifestyle choice. Many of us live in comfort and convenience provided by fossil fuels. If we've driven vehicles powered by internal combustion engines or heated our homes with wood, coal, oil, or gas, we've contributed to conditions that fueled the Los Angeles fires and the many extreme-weather disasters raging around the world.

So I circle back to the two questions that inspired this book:

1. What kind of world do we want for our children and theirs?
2. How must we manage affairs on spaceship Earth to achieve it?

Part III

Appendices

The purpose of information is not knowledge. It is being able to take the right action.

Peter Drucker

A

Earth in Real Time

Earth in 100 Seconds: Revealing Humanity's Global Footprint
https://www.youtube.com/watch?v=4yHz4WjWgtA

Human Population
https://www.theworldcounts.com/challenges/planet-earth/state-of-the-planet/world-population-clock-live

Atmospheric CO_2
https://www.co2.earth/daily-co2

Global Temperature
https://climate.nasa.gov/vital-signs/global-temperature/?intent=121

Global Fire Dashboard
https://maps.greenpeace.org/fire_dashboard/

Global Resource Extraction
https://www.theworldcounts.com

Global Income Inequality
https://wid.world/

Doomsday Clock
https://thebulletin.org/doomsday-clock/current-time/

Carbon Footprint Calculator
https://footprint.conservation.org/en-us

Environmental Footprint Calculator
https://www.footprintcalculator.org/home/en

Plastics Footprint Calculator
https://www.omnicalculator.com/ecology/plastic-footprint

Water Footprint Calculator
https://watercalculator.org

The IUCN Red List of Threatened Species
https:/www.iucnredlist.org/

Global Renewables Watch
https://www.globalrenewableswatch.org

—

Run experiments in the En-ROADS climate simulator to craft your own sustainability story.

En-ROADS — Choose your own future
https://en-roads.climateinteractive.org/scenario.html?v=25.10.0

B

Governing Frameworks

The United Nations 17 Goals
https://sdgs.un.org/goals

The United Nations Global Framework for Managing Nature Through 2030
https://www.un.org/sustainabledevelopment/blog/2021/07/a-new-global-framework-for-managing-nature-through-2030-1st-detailed-draft-agreement-debuts

Kunming-Montreal Global Biodiversity Framework
https://www.cbd.int/gbf

Agreement under the United Nations Convention on the Law of the Sea on the Conservation and Sustainable Use of Marine Biological Diversity of Areas beyond National Jurisdiction (BBNJ Agreement)
https://www.un.org/bbnjagreement/sites/default/files/2024-08/Text%20of%20the%20Agreement%20in%20English.pdf

Item 69 of the Constitution of Kenya
https://www.klrc.go.ke/index.php/constitution-of-kenya/116-chapter-five-land-and-environment/118-part-2-environment-and-natural-resources/236-69-obligations-in-respect-of-the-environment

United Nations Convention on the Law of the Sea
https://www.un.org/depts/los/convention_agreements/texts/unclos/unclos_e.pdf

C

Organizations

You can learn much about sustainability issues and concerns and who's doing what by studying the focus and initiatives of the following organizations. Note that descriptions are lightly edited from each organization's website.

Rated organizations

For ratings data and more, checkout the Charity Navigator website:

— https://www.charitynavigator.org

1% for the Planet

A global network with thousands of businesses and environmental organizations working together to support people and the planet.

— https://www.onepercentfortheplanet.org/

+350

We're an international movement of ordinary people working to end the age of fossil fuels and build a world of community-centered renewable energy for all.

— https://350.org/?r=US&c=NA

Audubon

Build a bright future for birds and the planet.

— https://www.audubon.org/

Berkeley Earth

Berkeley Earth empowers global decision-making on climate change by providing transparent, open-source environmental data and unbiased scientific analysis. Our mission is to deliver high-quality, accessible information on global temperature and air pollution to drive action and foster a deeper understanding of humanity's impact on the planet.

— https://berkeleyearth.org/

Carbon Underground

The Carbon Underground was created in 2013 with two core goals—coalesce the emerging science on soil's relationship to climate change and tell the world about a possibility to not simply slow down this threat, but reverse it.

— https://thecarbonunderground.org/

Climate Foundation

Dedicated to reversing global warming in our lifetime.

— https://www.climatefoundation.org/

Climate Reality Project

Our mission is to catalyze a global solution to the climate crisis by making urgent action a necessity across every sector of society.

— https://www.climaterealityproject.org/

Cool Effect

Our mission is to reduce carbon emissions.

— https://www.cooleffect.org/

Conservation International

Securing the health of Earth's climate, ecosystems and biodiversity is essential to the well-being of all people. Worldwide, Conservation International is working to improve the lives of people everywhere by protecting oceans, forests and other living ecosystems.

— https://www.conservation.org/

Cultural Survival

Cultural Survival advocates for Indigenous Peoples' rights and supports Indigenous communities' self-determination, cultures and political resilience since 1972.

— https://www.culturalsurvival.org/

Earth Guardians

With a global network spanning six continents, EG empowers youth worldwide to lead change in their communities, fostering a collaborative network to support their endeavors.

— https://www.earthguardians.org/

Earth Island Institute

Earth Island Institute provides support to environmental action projects and to the next generation of environmental leaders in order to achieve solutions to the crises threatening the survival of life on Earth.

— https://www.earthisland.org/

Earthjustice

Earthjustice is a nonprofit public interest environmental law organization. We wield the power of law and the strength of partnership to protect people's health, to preserve magnificent places and wildlife, to advance clean energy, and to combat climate change.

— https://earthjustice.org/

Earthwatch Institute

Earthwatch connects people with scientists worldwide to conduct environmental research and empowers them with the knowledge they need to conserve the planet.

— https://earthwatch.org

Earthworks

Ending oil & gas and mining pollution. We stand with front-line communities opposing fossil fuels and dirty mining while advocating for a clean energy transition.

— https://earthworks.org/

Ecological Landscape Alliance

Since its founding in 1992, the Ecological Landscape Alliance has been a leader in promoting sustainable approaches to landscape design, construction, and management. ELA's commitment to innovative ideas and evidence-based practices has made the organization both a trusted resource and a vibrant community of landscape professionals and devoted gardeners.

— https://www.ecolandscaping.org

Eden+

Human well-being and the health of the planet are inseparable + interdependent. By partnering with local communities to collaboratively design and implement landscape restoration projects, we are inspiring hope, enhancing biodiversity, and generating long-term socioeconomic benefits to strengthen the resilience of people and the planet.

— https://www.eden-plus.org/

Environmental Defense Fund

Climate change solutions. We deliver game-changing solutions that have a real impact for people around the globe. Our vision: A vital Earth. For everyone.

— https://www.edf.org/

Fibershed

We develop regional fiber systems that build soil & protect the health of our biosphere.

— https://fibershed.org/

Food & Water Watch

With more than 2 million supporters, Food & Water Watch fights for safe food, clean water, and a livable climate for all of us. We protect people from the corporations and other destructive economic interests that put profit ahead of everything else.

— https://www.foodandwaterwatch.org/

Friends of the Earth

Friends of the Earth fights for a more healthy and just world. We speak truth to power and expose those who endanger the health of people and the planet for corporate profit. We organize to build long-term political power and campaign to change the rules of our economic and political systems that create injustice and destroy nature.

— https://foe.org/

Global Footprint Network

Every year people demand more from nature than it can regenerate. Individuals, communities and government leaders use our data to better manage limited resources, reduce economic risk, and improve well-being.

— https://www.footprintnetwork.org/

Greenpeace

Greenpeace is a global network of independent campaigning organizations that use peaceful protest and creative confrontation to expose global environmental problems and promote solutions that are essential to a green, just, and joyful future.

— https://www.greenpeace.org/global/

Heifer International

Ending poverty begins with agriculture. Heifer International is on a mission to end hunger and poverty in a sustainable way by supporting and investing alongside local farmers and their communities.

— https://www.heifer.org/

Institute for Applied Ecology

Our vision is a world where all people and wildlands are healthy and interact positively, biological diversity flourishes, and environmental challenges are met with a social commitment to solving problems with scientific principles.

— https://appliedeco.org

International Renewable Energy Agency (IRENA)

International Renewable Energy Agency (IRENA) is an intergovernmental organization that supports countries in their transition to a sustainable energy future, and serves as the principal platform for international co-operation, a center of excellence, and a repository of policy, technology, resource and financial knowledge on renewable energy. IRENA promotes the widespread adoption and sustainable use of all forms of renewable energy, including bioenergy, geothermal, hydropower, ocean, solar and wind energy, in the pursuit of sustainable development, energy access, energy security and low-carbon economic growth and prosperity.

— https://www.irena.org/

Land Institute

We work for a future in which humans flourish as members of a thriving ecosphere. Achieving this future requires reconciling the human economy with nature's economy, and we believe focusing on food and how we produce it is a transformative first step.

— https://landinstitute.org

National Geographic Society

For more than 130 years, the National Geographic Society has funded the best and brightest individuals dedicated to scientific discovery and understanding of our world.

— https://www.nationalgeographic.org/society/

Natural Resources Defense Council

NRDC combines the power of more than 3 million members and online activists with the expertise of some 700 scientists, lawyers, and other environmental specialists to confront the climate crisis, protect the planet's wildlife and wild places, and to ensure the rights of all people to clean air, clean water, and healthy communities.

— https://www.nrdc.org/

Nature Conservancy

The Nature Conservancy is tackling the dual threats of accelerated climate change and unprecedented biodiversity loss. Science determines where we focus and equity guides how we achieve lasting results. Grounded by decades of local on-the-ground experience, we maximize our ability to affect change by bringing together real-world solutions, policy expertise, sustainable financing and collaborative partnerships.

— https://www.nature.org/en-us/

Ocean Cleanup

Our cleaning technologies are deployed around the world as we conduct the largest cleanup in history. For over ten years, The Ocean Cleanup has been researching, extracting, and monitoring plastic pollution in oceans and rivers globally – with millions of kilograms removed to date.

— https://theoceancleanup.com/

Ocean Conservancy

Ocean Conservancy works to protect the ocean from today's greatest global challenges. We create evidence-based solutions for a healthy ocean and the wildlife and communities that depend on it.

— https://oceanconservancy.org/

Oceana

Oceana is the largest international advocacy organization focused solely on ocean conservation. Oceana's mission is to protect and restore our oceans.

— https://oceana.org/

Omega Institute for Holistic Studies

Omega is a nonprofit, donor-supported, educational organization at the forefront of holistic studies. Located on more than 250 acres in New York's beautiful Hudson Valley, Omega offers a wide variety of workshops, retreats, conferences, and professional training, in-person and online. Since 1977, more than one million people have come to Omega for reasons big and small—to spark creativity, explore spirituality, improve well-being, and connect to a community of lifelong learners.

— https://www.eomega.org/

One Tree Planted

We want to make it simple for anyone to help the environment by planting trees. Together we can restore forests, create habitat for biodiversity, and make a positive social impact around the world.

— https://onetreeplanted.org/

Only One

We're on a mission to restore ocean health and tackle the climate crisis in this generation.

— https://only.one/

Oxfam

We are a global organization that fights inequality to end poverty and injustice.

— https://www.oxfamamerica.org

Penn State Institute of Energy and the Environment

The Institute of Energy and the Environment (IEE) connects and supports interdisciplinary teams of researchers that are solving some of the world's most difficult energy and environmental challenges.

— https://iee.psu.edu/home

Post Carbon Institute

Founded in 2003, Post Carbon Institute's mission is to lead the transition to a more resilient, equitable, and sustainable world by providing individuals and communities with the resources needed to understand and respond to the interrelated ecological, economic, energy, and equity crises of the 21st century.

— https://www.postcarbon.org/

Project Drawdown

Project Drawdown is an independent organization driving meaningful climage action by connecting people to science-based climate climate solutions and strategies.

— https://drawdown.org/

Rainforest Alliance

The Rainforest Alliance is an international non-profit organization working at the intersection of business, agriculture, and forests to make responsible business the new normal. We are building an alliance to protect forests, improve the livelihoods of farmers and forest communities, promote their human rights, and help them mitigate and adapt to the climate crisis.

— https://www.rainforest-alliance.org/

Regeneration International

Our mission: To promote, facilitate and accelerate the global transition to regenerative food, farming and land management for the purpose of restoring climate stability, ending world hunger and rebuilding deteriorated social, ecological and economic systems.

— https://regenerationinternational.org/

Restore the Earth Foundation

Our mission is to restore the Earth's essential forest and wetland ecosystems. We envision the Earth in balance — its original vitality and natural abundance available to all, for generations to come.

— https://restoretheearth.org/

Save the Redwoods League

Save the Redwoods League envisions vibrant redwood forests of the scale and grandeur that onece graced the California coast and the Sierra Nevada, protected forever, restored to grow old again, and connected to people through a network of magnificent parks and protected areas that inspire all of us with the beauty and power of nature.

— https://www.savetheredwoods.org/

Sierra Club

The Sierra Club is the most enduring and influential grassroots environmental organization in the United States. We amplify the power of our millions of members and supporters to defend everyone's right to a healthy world.

— https://www.sierraclub.org/

Sunrise Movement Educational Fund

Our objective: We are the climate revolution

Together, we will force the government to end the era of fossil fuel elites, invest in Black, brown and working class communities, and create millions of good union jobs. We're on a mission to put everyday people back in charge and build a world that works for all of us, now and for generations to come.

— https://www.sunrisemovement.org/

Sustainable Harvest International

Working to create a just and sustainable world through transformative farmer training that nourishes communities and the Earth.

— https://www.sustainableharvest.org/

Waterkeeper Alliance

We're a global movement of community-based advocates united for clean, healthy, and abundant water for all people and the planet. We preserve and protect water by connecting local Waterkeeper groups worldwide.

— https://waterkeeper.org/

World Resources Institute

We envision a new era in which the world can meet people's needs while curbing climate change and protecting nature. We see a future where everyone can enjoy clean air, walkable cities, vibrant landscapes, nutritious food and affordable energy. Ecosystems are thriving. The climate is stable.

— https://www.wri.org/

World Wildlife Fund

For more than 60 years, WWF has worked to help people and nature thrive. As the world's leading conservation organization, WWF works in nearly 100 countries. At every level, we collaborate with people around the world to develop and deliver innovative solutions that protect communities, wildlife, and the places in which they live.

— https://www.worldwildlife.org/

Unrated organizations

5GYRES

Marcus Eriksen and Anna Cummins founded the 5 Gyres Institute to investigate key unanswered questions about plastic pollution.

— https://www.5gyres.org/

A Growing Culture

Food sovereignty is a human right. Healthy, accessible, and culturally appropriate food is a human right. Living wages, living environments, living traditions are human rights. Dignity. Diversity. Culture. All human rights. These are the bedrock of the food sovereignty movement. We believe in, and we will stop at nothing until there is food sovereignty. For everyone. Everywhere.

— https://www.agrowingculture.org/

American Ecological Engineering Society

Ecological engineering, defined as the design of sustainable ecosystems that integrate human society with its natural environment for the benefit of both, has developed over the last 30 years, and rapidly over the last 10 years. Its goals include the restoration of ecosystems that have been substantially disturbed by human activities and the development of new sustainable ecosystems that have both human and ecological values.

— https://www.ecoeng.org

American Forests

Since our founding in 1875, we have been the pathfinders for creating healthy forests from coast to coast. For example, we championed the creation of the US Forest Service and have persuaded Congress to provide stable funding for fighting and preventing forest fires. Our deep knowledge of forests and track record of collaboration position us to build a reforestation movement in America.

— https://www.americanforests.org/

Cascade Institute

A Canadian research center addressing urgent and entangled global problems. We anticipate future crises and identify high-leverage interventions to change humanity's course for the better.

— https://cascadeinstitute.org/

Center for Ecological Design

Mission: To create an inclusive laboratory for discovery and experimentation with ecological design: grassroots art and technology that cultivate social and environmental well-being.

— https://www.centerforecologicaldesign.org/technology/ecological-design

Depave

Depave empowers disenfranchised communities to overcome social and environmental injustices and adapt to climate change through urban re-greening.

— https://www.depave.org/

Extinction Rebellion

Extinction Rebellion is a global nonviolent movement to compel the world to address the climate and ecological emergency.

– https://extinctionrebellion.us/

Fairtrade International

Fairtrade changes the way trade works through better prices, decent working conditions and a fairer deal for farmers and workers in developing countries.

— https://www.fairtrade.net/en.html

Farming Falmouth

Our mission is to build a stronger food system for our town— one that nurtures personal health, community health, and the health of our planet. We work to improve access to nutrition for all.

— https://www.farmingfalmouth.org/

Fossil Free Media

Fossil Free Media is a nonprofit media lab that supports the movement to end fossil fuels and address the climate emergency.

— https://fossilfree.media/#welcome

Global Institute for Water Security

Our Vision is a world where water is managed equitably and sustainably for protection and enhanced quality of life, strong economies, and healthy ecosystems.

— https://water.usask.ca/

Green Center

The Green Center is a non-profit educational institute that evolved from The New Alchemy Institute. With the help of a small board of directors and volunteers, former New Alchemists Hilda Maingay and Earle Barnhart continue NAI's mission to "create ecologically-derived human support systems" which include renewable energy, agriculture, aquaculture, housing, and landscapes.

— https://newalchemists.net/

Green Cross

Green Cross works to prevent and resolve conflicts arising from environmental degradation and the growing demand for increasingly scarce natural resources with a focus on addressing the global water and energy challenges.

— https://www.gcint.org/

Jane Goodall Institute

Dr. Jane Goodall discovered that when we put local communities at the heart of conservation, we improve the lives of people, animals and the environment. JGI advances Dr. Goodall's holistic approach through a tapestry of nine strategies that build on each other and bring the power of community-centered conservation to life.

— https://janegoodall.org/

Lonely Whale

Lonely Whale has spearheaded impactful global movements to raise awareness and offer alternatives to problematic plastics, convened a consortium of multi-national brands to create the first global network of ocean-bound plastic supply chains, and supported the global youth movement with the tools and network needed to advance their efforts to secure a future with clean seas.

— https://www.lonelywhale.org/

NĀTIFS: North American Traditional Indigenous Food Systems

NATIFS works to cultivate the seeds of Indigenous heritage, nourish our people, and honor the land that sustains us.

— https://natifs.org/

Native Seeds/SEARCH

For more than four decades, Native Seeds/SEARCH (NS/S) has stewarded the seeds of the desert Southwest and Mexico. Founded in 1983 in response to the concern of farmers, gardeners, Indigenous community members, and conservationists about the devastating loss of seed diversity, NS/S now conserves more than 1,800 regional seed varieties in a climate-controlled seed bank.

— https://www.nativeseeds.org

Omega Center for Sustainable Living

Operating since 2009, the Omega Center for Sustainable Living (OCSL) integrates our award-winning building and the campus's state-of-the-art natural water reclamation facility—the Eco Machine™—with a robust environmental education program, focused on a whole-systems, place-based approach to problem solving that considers the connections between society, nature, and individuals.

— https://www.eomega.org/center-sustainable-living/

Pacific Institute

The Pacific Institute creates and advances solutions to the world's most pressing water challenges. Our vision is to create a world in which society, the economy, and the environment have the water they need to thrive now and in the future.

— https://pacinst.org/

Planet Drum Foundation

Planet Drum was founded in 1973 in San Francisco, California to provide an effective grassroots approach to ecology that emphasizes sustainability, community self-determination and regional self-reliance. In association with community activists and ecologists, Planet Drum developed the concept of a bioregion. Through its projects, publications, speakers, and workshops, Planet Drum helps start new bioregional groups and encourages local organizations and individuals to find sustainable ways to live within the natural confines of bioregions. We believe that people who know and care about the places where they live will work to maintain and restore them.

— https://planetdrum.org/

Regenerative Agriculture Alliance

Regenerative Agriculture Alliance was created in 2018 as the non-profit arm of a larger ecosystem designed to build a regenerative, equitable, and socially just agriculture sector. We are focused on scaling up a systems-level regenerative poultry solution that restores ecological balance, produces nourishing food, and puts money back into the hands of farmers and food chain workers.

— https://www.regenagalliance.org/

Resilience

Resilience aims to support building community resilience in a world of multiple emerging challenges: the decline of cheap energy, the depletion of critical resources like water, complex environmental crises like climate change and biodiversity loss, and the social and economic issues which are linked to these.

— https://www.resilience.org/

SeaLegacy

SeaLegacy's mission is to inspire people to fall in love with the ocean, amplify a network of change makers around the world, and catalyze hands-on diplomacy through hopeful, world-class visual storytelling.

— https://www.sealegacy.org/

Soil and Water Conservation Society

An international organization for professionals who practice and advance the science and art of natural resource conservation.

— https://www.swcs.org/about-us/our-organization

Soil Association

Soil Association is working to transform the way we eat, farm and care for our natural world.

— https://www.soilassociation.org/

Third Act

We're a community of Americans over sixty determined to change the world for the better. Third Act harnesses an unparalleled generational power to safeguard our climate and democracy.

— https://thirdact.org/

WeForest

We work in threatened and degraded tropical forests and landscapes to create locally-led reforestation projects.

— https://www.weforest.org/

World Health Organization Sustainable Development

Dedicated to the well-being of all people and guided by science, the World Health Organization leads and champions global efforts to give everyone, everywhere an equal chance to live a healthy life.

— https://www.who.int/health-topics/sustainable-development\

D

Publications

Newsletters

Wondering what it might be like to live in a sustainable doughnut world? These general interest publications are chock-a-block with ideas and specific things you can do to evolve toward a sustainable lifestyle.

TreeHugger

— https://www.treehugger.com/

The Ethicalist

— https://theethicalist.com/

Sustainability Times

— https://www.sustainability-times.com/

Research Guides

— https://guides.library.ucla.edu/c.php?g=180477\&p=1191364

— https://www.scimagojr.com/index.php

E

Books

Given the inescapable facts of climate change and geopolitical conflict, long-term sustainability of human life on Earth is an urgent matter. But the concept is abstract— scary even, easy to dismiss under the pressures of daily concerns. But if we have, or hope to have, children, or if we careabout the vitality of the natural world, we have a vital stake in the matter. Day-to-day decisions and conmittments of consumers, business executives, and governing policy makers set the course toward sustainability or civilizational collapse. That means us. Each of us. We each bear responsiblity for the long-term future of humanity on Earth whether we care to accept it or not.

But what to do? That's the question.

The first step is to understand what's at stake, within what time constraints, and who's doing what to address the issue.

Enter "sustainability + books" into your browser and you'll find enough titles to keep yourself immersed in the topic for months if not more. Books range from general-reader popularizations to business strategies for corporate managers to technical books for specialists.

Kate Raworth's *Doughnut Economics: Seven Ways to Think Like a 21st-Century Economist* provides a pragmatic perspective and framework for organizing thought and action. I recommend it as a must read.

But set out to learn more about sustainability and you'll quickly realize that the concept is rather fuzzy. The UCLA library suggests more than 160 search terms you might use to delve into the topic. We must ask, what is being sustained? Why does it matter? And how will we know when we've succeeded? Merriam

Webster offers this: "...relating to, or being a method of harvesting or using a resource so that the resource is not depleted or permanently damaged." My interests, however, are more encompassing.

Motivated by ever more extreme weather events and the progressively more alarming reports from the Intergovernmental Panel on Climate Change (IPCC), I became obsessed with the sustainability of humanity on Earth given the challenges and short window of opportunity imposed by global heating and ravaged ecosystems.

IPCC: "Some climate impacts are already so severe they cannot be adapted to, leading to losses and damages." As a non-specialist I quickly learned that I'd entered a thorny maze.

You may find the following bibliography rather quirky. But it marks way points, breadcrumbs if you will, that trace my missteps and turns through the maze. I'd read *Limits to Growth* and *Small is Beautiful* in the early '70s and they'd lingered in my mind as faint but gnawing anxieties. *America Beyond Capitalism* and *Beyond Growth* lured me well into the maze. *Regeneration* was a crucial turn. Every book in the list has contributed insights and mini-ahas along the twisty path.

——-

Aklin, M. (2018). *Renewables: the Politics of a Global Energy Transition.* MIT Press, Cambridge, MA

Alperovitz, G. and Speth, J. G. (2011) *America Beyond Capitalism: Reclaiming our Wealth, Our Liberty, and Our Democracy.* Democracy Collaborative Press, Boston, MA

Ballou, B. (2023) *Plunder: Private Equity's Plan to Pilage America.* PublicAffairs, New York, NY

Bardi, U. and Pereira, C. A. (2022) *Limits and Beyond: 50 Years on from The Limits to Growth, What Did We Learn and What's Next?* Expat Presa, Sheridan, WY

Bierman, P. (2024) *When the Ice is Gone.* Simon & Norton, New York, NY

Brown D.R. and Agosta, S.J (2024) *A Darwinian Survival Guide: Hope for the Twenty-First Century.* MIT Press, Cambridge, MA

Cashman, G. (2013) *What Causes War? An Introduction to Theories of International Conflict.* Rowman & Littlefield Publishers, New York, NY

Daly, H.E., Cobb, J.B. and Cobb, C.W. (1994) *For the Common Good: Redirecting the Economy Toward Community, the Environment, and a Sustainable Future.* Beacon Press, Boston, MA

Daly, H.E. (1996) *Beyond Growth: the Economics of Sustainable Development.* Beacon Press, Boston, MA

de Schutter, O. (2024) *The Poverty of Growth.* Pluto Press, London, UK.

Deaton, A. (2024) *Economics in America.* Princeton University Press, Princeton, NJ

DiResta, R. (2024) *Invisible Rulers.* PublicAffairs, New York, NY

Hawken, P. (2021) *Regeneration: Ending the Climate Crisis in One Generation.* Penguin Books, New York, NY

Hawden, P., Lovins, P., Lovins, L.H. (1999) *Natural Capitalism: Creating the Next Industrial Revolution.* Little Brown and Company, Boston, MA

Heinberg, R. (2021) *Power: Limits and Prospects for Human Survival.* New Society Publishers, Gabriola Island, BC

Hoeschele, W. (2016) *The Economics of Abundance.* CRC Press, Abington, UK

Jackson, T. (2021) *Post Growth: Life after Capitalism.* Polity Press, Cambridge, UK

Kimmerer, R. W. (2024) *The Serviceberry: Abundance and Reciprocity in the Natural World.* Scribner, New York, NY

Lloyd, K. (2025) *Intra Terrestrials: Discovering the Strangest Life on Earth.* Princeton University Press, Princeton, NJ

McCoy, Alfred W. (2021) *To Govern the Globe: World Orders & Catastrophic Change.* Haymarket Books, Chicago, IL

Meadows, D.H., Meadows, D.L., Randers, J., Behrens III, W.W (1972) *The Limits to Growth: A Report for the Club of Rome's Project on the Predicament of Mankind.* Universe Books, New York, NY

Piketty, T. (2023) *Nature, Culture, and Inequality: A Comparative and Historical Perspective.* Other Press, New York, NY

Polanyi, K. (1944) *The Great Transformation: The Political and Economic Origins of Our Time.* Beacon Press, Boston, MA

Raworth, K. (2017) *Doughnut Economics: Seven Ways to Think Like a 21st-Century Economist.* Random House Business Books, London, UK

Ruskin, J. (1912) *Unto This Last.* available as Project Gutenberg eBook https://www.gutenberg.org/ebooks/36541

Schumacher, E.F. (1973) *Small is Beautiful: Economics as if People Mattered.* Harper Perennial, New York, NY

Smil, V. (2013) *Harvesting the Biosphere: What We Have Taken from Nature.* MIT Press, Cambridge, MA

Tainter, J.A. (1988) *The Collapse of Complex Societies.* Cambridge University Press, Cambridge, UK

Temin, P. (2018) *The Vanishing Middle Class: Prejudice and Power in a Dual Economy.* The MIT Press, Cambridge, MA

Todd, N.J. (2005) *A Safe and Sustainable World: The Promise of Ecological Design.* Island Press, Washington, DC

Todd, N.J. and Todd, J. (1994) *From Eco-Cities to Living Machines: Principles in Ecological Design.* North Atlantic Books, Berkeley, CA

University of California at Los Angles (2024) *Research Guides: Sustainability: Sustainability Topics.* Available at: https://guides.library.ucla.edu/c.php?g=180477\&p=1191364

Van Der Ryn, S. (1996) *Ecological Design.* Island Press, Washington, DC

Varoufakis, V. (2024). *Technofeudalism: What Killed Capitalism.* Melville House, Brooklyn, NY

Vaughn-Lee, L. ed. (2023) *Spiritual Ecology: The Cry of the Earth, Second Ed.* The Golden Sufi Center, Point Reyes, CA

von Weizsacker, E.U. (1998) *Factor Four: Doubling Wealth, Halving Resource Use — A Report to the Club of Rome.* Earthscan, London, UK

Notes

[1] NASA Global Climate Change. *Global Surface Temperature — NASA Global Climate Change*. 2024. URL: https://climate.nasa.gov/vital-signs/global-temperature/?intent=121.

[2] Sigred Lind, Randi B Ingvaldsen, and Tores Furek. *Arctic warming hotspot in the northern Barents Sea linked to declining sea-ice import*. 2018. URL: https://www.nature.com/articles/s41558-018-0205-y.

[3] Mika Rantanen et al. *The Arctic has warmed nearly four times faster than the globe since 1979*. 2022. URL: https://www.nature.com/articles/s43247-022-00498-3.

[4] Melissa B Mahle. *Great Power Competition in the Arctic: Implications for International Trade*. 2024. URL: https://www.steptoe.com/en/news-publications/stepwise-risk-outlook/great-power-competition-in-the-arctic-implications-for-international-trade.html.

[5] Jan Hjort et al. *Degrading permafrost puts Arctic infrastructure at risk by mid-century*. 2018. URL: https://www.nature.com/articles/s41467-018-07557-4?_ga=2.3180275.2125967820.1592296737-1913233615.1592296737.

[6] Hanna C. *Melting Permafrost Could Mean Return of Ancient Diseases*. 2020. URL: https://www.sciencetimes.com/articles/26903/20200817/melting-permafrost-mean-return-ancient-diseases.htm.

[7] W. J. Ripple et al. *The 2023 state of the climate report: Entering uncharted territory*. 2023. URL: https://academic.oup.com/bioscience/article/73/12/841/7319571.

[8] dlewis. *Highest number of countries engaged in conflict since World War II.* 2024. URL: https://www.visionofhumanity.org/highest-number-of-countries-engaged-in-conflict-since-world-war-ii/%5C#:~:text=fear%5C%20of%5C%.

[9] R. Pallardy. *List of wars.* 2015. URL: https://www.britannica.com/topic/list-of-wars-2031197.

[10] Suma K. *How Climate Change Drives Conflict and War Crimes Around the Globe.* 2023. URL: https://insideclimatenews.org/news/26102023/how-climate-change-drives-conflict-and-war-crimes-around-the-globe/.

[11] MHN Military History Now. *First Blood — History's Earliest Recorded Military Conflicts.* 2014. URL: https://militaryhistorynow.com/2014/10/13/first-blood-historys-earliest-recorded-military-conflicts/%5C#google%5C_vignette.

[12] R. Jackson. *Who Discovered the Greenhouse effect?* 2019. URL: https://www.rigb.org/explore-science/explore/blog/who-discovered-greenhouse-effect.

[13] Anon. 2019. URL: https://publicdomainreview.org/collection/first-paper-to-link-co2-and-global-warming-by-eunice-foote-1856/.

[14] Quanta Magazine. *Physicists Pinpoint the Quantum Origin of the Greenhouse Effect, Quanta Magazine.* 2024. URL: https://www.quantamagazine.org/physicists-pinpoint-the-quantum-origin-of-the-greenhouse-effect-20240807/.

[15] Wikimedia Commons. *File:20200324 Global average temperature - NASA-GISS HadCrut NOAA Japan BerkeleyE.svg.* 2025. URL: https://commons.wikimedia.org/wiki/File:20200324%5C_Global%5C_average%5C_temperature%5C_-%5C_NASA-GISS%5C_HadCrut%5C_NOAA%5C_Japan%5C_BerkeleyE.svg.

[16] Giacomo Falchetta et al. *Global projections of heat exposure of older adults.* 2024. URL: https://www.nature.com/articles/s41467-024-47197-5.

[17] E. Robinson and R.C. Robbins. *Smoke & Fumes.* 1968. URL: https://www.smokeandfumes.org/documents/16.

[18] G. Dembicki. *US oil company ran 1977 article predicting climate crisis could cause starvation.* 2024. URL: https://www.theguardian.com/us-news/article/2024/jul/18/us-oil-marathon-petroleum-climate-change.

[19] M. Copley. *Exxon minimized climate change internally after conceding that fossil fuels cause it.* 2023. URL: https://insideclimatenews.org/news/26102023/how-climate-change-drives-conflict-and-war-crimes-around-the-globe/.

[20] Elliot Negin. *ExxonMobil Claims Shift on Climate But Continues to Fund Climate Science Deniers*. 2020. URL: https://blog.ucs.org/elliott-negin/exxonmobil-claims-shift-on-climate-continues-to-fund-climate-deniers/.

[21] Joint Staff Report. *Denial, Disinformation, and Doublespeak: Big Oil's Evolving Efforts to Avoid Accountability for Climate Change*. 2024. URL: https://www.budget.senate.gov/imo/media/doc/fossil%5C_fuel%5C_report1.pdf.

[22] Chuck Collins, Helen Fannery, and Bella DeVaan. *Inside the World of Fossil Fuel Philanthropy*. 2024. URL: https://inequality.org/article/fossil-fuel-philanthropy/.

[23] climate.mit.edu. *The 1.5 C climate benchmark*. 2023. URL: https://climate.mit.edu/posts/explained-15-c-climate-benchmark.

[24] M. Poynting. *World's first year-long breach of key 1.5C warming limit*. 2024. URL: https://www.bbc.com/news/science-environment-68110310.

[25] M. Lempriere. *Wind and solar are 'fastest-growing electricity sources in history'*. 2024. URL: https://www.carbonbrief.org/wind-and-solar-are-fastest-growing-electricity-sources-in-history/.

[26] M. Roser. *Why did renewables become so cheap so fast? Our World in Data*. 2020. URL: https://ourworldindata.org/cheap-renewables-growth.

[27] Matt Solomon. *The Rise of Batteries in Six Charts and Not Too Many Numbersg*. 204. URL: https://rmi.org/the-rise-of-batteries-in-six-charts-and-not-too-many-numbers/.

[28] ICAN. *Surge: 2023 Global nuclear weapons spending*. 2023. URL: https://www.icanw.org/surge%5C_2023%5C_global%5C_nuclear%5C_weapons%5C_spending.

[29] J. Mecklin. *A Time of Unprecedented Danger: It Is 90 Seconds to Midnight. Bulletin of the Atomic Scientists*. 2024. URL: https://thebulletin.org/doomsday-clock/current-time/.

[30] Kate Raworth. *About Kate Raworth*. 2025. URL: https://www.kateraworth.com/about/.

[31] Worldometer. *World Population by Year - Worldometers*. 2023. URL: https://www.worldometers.info/world-population/world-population-by-year/.

[32] *Deed - Attribution-ShareAlike 4.0 International - Creative Commons*. URL: https://creativecommons.org/licenses/by-sa/4.0/deed.en.

[33] Anshool Deshmukh. *The Global Wealth Distribution*. https://www.visualcapitalist.com/distribution-of-global-wealth-chart/. 2021.

[34] Martínez . A and Obed Manuel. *More billionaires than ever ranked in Forbes' annual list. Here are the top 10*. 2025. URL: https://www.npr.org/2025/04/01/nx-s1-5345950/forbes-billionaires-list.

[35] United Nations UNifeed. *FAO / FOOD SECURITY REPORT*. 2024. URL: https://media.un.org/unifeed/en/asset/d323/d3239510.

[36] World Health Organization. *Joint child malnutrition estimates (JME)*. 2023. URL: https://www.who.int/teams/nutrition-and-food-safety/monitoring-nutritional-status-and-food-safety-and-events/joint-child-malnutrition-estimates.

[37] UN News. *UN searches for solutions to global housing crisis*. 2025. URL: https://news.un.org/en/story/2025/05/1163851.

[38] Tanya de Sousa and Meghan Henry. 2024. URL: https://www.huduser.gov/portal/sites/default/files/pdf/2024-AHAR-Part-1.pdf.

[39] CDC. *About Homelessness and Health*. 2024. URL: https://www.cdc.gov/homelessness-and-health/about/index.html.

[40] Z. Lin, E. Weinberger, and et al. Nori-Sarma N. *Daily heat and mortality among people experiencing homelesness in 2 urban US counties, 2015-2022*. 2024. URL: https://academic.oup.com/aje/article/193/11/1576/7689061.

[41] Summer Lin. *'He baked': Heat waves are killing more L.A. homeless people who can't escape broiling sun*. 2023. URL: https://www.latimes.com/california/story/2023-02-19/la-me-homeless-heat-deaths.

[42] AP News. *Divided Supreme Court rules in major homelessness case that outdoor sleeping bans are OK*. 2024. URL: https://apnews.com/article/supreme-court-homeless-camping-bans-506ac68dc069e3bf456c10fcedfa6bee.

[43] A. Johnson. *Advocating internment camps for homeless people is increasingly mainstream*. 2023. URL: https://therealnews.com/advocating-internment-camps-for-homeless-people-is-increasingly-mainstream-this-should-frighten-us..

[44] data.unhcr.org. *Situation Ukraine Refugee Situation*. 2024. URL: https://data.unhcr.org/en/situations/ukraine/location/680.

[45] Soph Warnes et al. *Gaza after 15 months of war: A snapshot in maps and charts*. 2025. URL: https://www.cnn.com/2025/01/22/middleeast/gaza-war-ceasefire-snapshot-dg-intl.

[46] OCHA. *Sudan. reports.unocha.org*. 2024. URL: https://reports.unocha.org/en/country/sudan/.

[47] Al Jazeera. *Bleak milestone: UN says 3 million forced to flee in Myanmar conflict*. 2024. URL: https://www.aljazeera.com/news/2024/5/8/bleak-milestone-un-says-3-million-forced-to-flee-in-myanmar-conflict.

[48] V. Clement et al. *Groundswell Part 2. openknowledge.worldbank.org*, 2. 2021. URL: https://openknowledge.worldbank.org/entities/publication/2c9150df-52c3-58ed-9075-d78ea56c3267.

[49] Berners-Lee M. et al. *Current global food production is sufficient to meet human nutritional needs in provided there is radical societal adaptation*. 2024. URL: https://online.ucpress.edu/elementa/article/doi/10.1525/elementa.310/112838/Current-global-food-production-is-sufficient-to.

[50] L. Gutierrez. *Housing for All: Addressing the Social and Environmental Implications of the Housing Crisis - Sustainable Earth*. 2023. URL: https://sustainable-earth.org/housing-crisis/.

[51] GBD 2021 Fertility and Forecasting Collaborators. *Global fertility in 204 countries and territories, 1950—2021, with forecasts to 2021: a comprehensive demographic analysis for the Global Burden of Disease Study 2021*. 2024. URL: https://www.thelancet.com/journals/lancet/article/PIIS0140-6736(24)00550-6/fulltext.

[52] World Population Review. *Birth Rate by Country 2025*. 2025. URL: https://worldpopulationreview.com/country-rankings/birth-rate-by-country.

[53] *World Population by Country in 2065 (World Map)*. 2024. URL: https://database.earth/population/by-country/2065.

[54] D. MacRae. *Britannica Money*. 2024. URL: https://www.britannica.com/money/Thomas-Malthus.

[55] S. Hurt. *Washington Consensus | History*. 2025. URL: https://www.britannica.com/money/Washington-consensus.

[56] Real Country Sizes. *Pakistan Real Size: Bigger Than Texas, 2x Bigger Than Japanal Country Sizes — realcountrysizes.com*. URL: https://realcountrysizes.com/true-size-of-pakistan/.

[57] Wikipedia. *2022 Pakistan floods*. URL: https://en.wikipedia.org/wiki/2022%5C_Pakistan%5C_floods.

[58] Dawn. *As 2022 comes to a close, millions of Pakistanis still reel from catastrophic floods*. 2022. URL: https://www.dawn.com/news/1724134.

[59] World Bank Group. *Pakistan: Flood Damages and Economic Losses Over USD 30 billion and Reconstruction Needs Over USD 16 billion*. 2022. URL: https://www.worldbank.org/en/news/press-release/2022/10/28/pakistan-flood-damages-and-economic-losses-over-usd-30-billion-and-reconstruction-needs-over-usd-16-billion-new-assessme.

[60] UNICEF. *Devastating Floods in Pakistan*. 2023. URL: https://www.unicef.org/emergencies/devastating-floods-pakistan-2022.

[61] Ian Hall. *The Canadian wildfires have now burned boreal forests...* 2024. URL: https://x.com/IanHall_CU/status/1698233885883060618.

[62] K. Viner. *Canada's 2023 wildfires released more greenhouse gases than most countries. the Guardian*. 2024. URL: https://www.theguardian.com/world/article/2024/aug/29/canada-wildfires-carbon-emissions.

[63] J. Christensen. *It's so hot in Arizona, doctors are treating a spike of patients who were burned by falling on the ground*. 2023. URL: https://www.cnn.com/2023/07/24/health/arizona-heat-burns-er/index.html.

[64] L. Taylor. *We can't keep up*. 2023. URL: https://www.theguardian.com/world/2023/sep/06/panama-colombia-border-migrants-record.

[65] public.wmo.int. *Earth had hottest three-month period on record, with unprecedented sea surface temperatures and much extreme weather*. 2023. URL: https://public.wmo.int/en/media/press-release/earth-had-hottest-three-month-period-record-unprecedented-sea-surface.

[66] W. Zhi et al. *Widespread deoxygenation in warming rivers*. 2023. URL: https://www.nature.com/articles/s41558-023-01793-3.

[67] C. H. Jaynes. *US sets record for billion-dollar disasters in a single year, with almost four months to go*. 2023. URL: https://www.nationofchange.org/2023/09/15/u-s-sets-record-for-billion-dollar-disasters-in-a-single-year-with-almost-four-months-to-go/.

[68] R. Lindsey. *Climate Change: Atmospheric Carbon Dioxide*. 2024. URL: https://www.climate.gov/news-features/understanding-climate/climate-change-atmospheric-carbon-dioxide.

[69] A. Mukpo. *Nitrogen: The environmental crisis you haven't heard of yet*. 2021. URL: https://news.mongabay.com/2021/09/nitrogen-the-environmental-crisis-you-havent-heard-of-yet/.

[70] B. Strauss. *10 Reasons Animals Go Extinct*. 2016. URL: https://www.thoughtco.com/reasons-animals-go-extinct-3889931.

[71] C. Llor and L. Bjerrum. *Antimicrobial resistance: risk associated with antibiotic overuse and initiatives to reduce the problem*. 2014. URL: https://journals.sagepub.com/doi/10.1177/2042098614554919.

[72] Club of Rome. *Homepage*. URL: https://www.clubofrome.org/.

[73] D. H. Meadows et al. *The Limits to Growth: A report for the Club of Rome's project on the predicament of mankind*. 1972. URL: https://www.clubofrome.org/publication/the-limits-to-growth/.

[74] U. Bardi and Carlos Alvarez Pereira. *Limits and Beyond*. Expat Press, 2022. URL: https://exapt.press/books/limits-and-beyond.

[75] D. Marchese. *This Pioneering Economist Says Our Obsession With Growth Must End*. 2022. URL: https://www.nytimes.com/interactive/2022/07/18/magazine/herman-daly-interview.html.

[76] K. Raworth. *Doughnut economics: seven ways to think like a 21st-century economist*. New York, N.Y.: Chelsea Green Publishing, 2017.

[77] F. Harvey. *'Massive disinformation campaign' is slowing global transition to green energy. the Guardian*. 2024. URL: https://www.theguardian.com/environment/article/2024/aug/08/fossil-fuel-industry-using-disinformation-campaign-to-slow-green-transition-says-un.

[78] D. Robertson. *How 'Owning the Libs' Became the GOP's Core Belief*. 2021. URL: https://www.politico.com/news/magazine/2021/03/21/owning-the-libs-history-trump-politics-pop-culture-477203.

[79] A. Smith. *2023: A historic year of U. S. billion-dollar weather and climate disasters | NOAA Climate.gov. www.climate.gov*. 2024. URL: https://www.climate.gov/news-features/blogs/beyond-data/2023-historic-year-us-billion-dollar-weather-and-climate-disasters.

[80] K. C. Rose et al. *Aquatic deoxygenation as a planetary boundary and key regulator of Earth system stability*. 2024. URL: https://www.nature.com/articles/s41559-024-02448-y.

[81] A.C. Matear R.J. an Hirst. *Long-term changes in dissolved oxygen concentrations in the ocean caused by protracted global warming*. 2003. URL: https://agupubs.onlinelibrary.wiley.com/doi/full/10.1029/2002GB001997.

[82] M. Igini. *The Tipping Points of Climage Change: How will Our World Change?* 2024. URL: https://earth.org/tipping-points-of-climate-change/.

[83] Gupta J. Johan Rockström et al. *Safe and just Earth system boundaries.* 2023. URL: https://www.nature.com/articles/s41586-023-06083-8.

[84] D. Gayle. *Earth may have breached seven of nine planetary boundaries, health check shows. the Guardian.* 2024. URL: https://www.theguardian.com/environment/2024/sep/23/earth-breach-planetary-boundaries-health-check-oceans.

[85] www.merriam-webster.com. *Definition of ABUNDANCE.* URL: https://www.merriam-webster.com/dictionary/abundance.

[86] J. Godoy. *FTC Looking Into Targeted Pricing Based on Personal Data.* 2024. URL: https://www.insurancejournal.com/news/national/2024/07/26/785669.htm.

[87] J. Hasell et al. *Economic Inequality. Our World in Data.* 2023. URL: https://ourworldindata.org/economic-inequality.

[88] C. Munoz and P. Anthem. *11 facts about food loss and waste – and how it links to sustainable food systems.* 2021. URL: https://www.wfp.org/stories/11-facts-about-food-loss-and-waste-and-how-it-links-sustainable-food-systems.

[89] W. Hoeschele. *The Economics of Abundance.* Boca Raton, FL, USA: CRC Press, 2016.

[90] E. Schlosser. *Do We Really Want a Food Cartel?* 2024. URL: https://www.theatlantic.com/ideas/archive/2024/04/food-industry-monopoly-power/678005/.

[91] M. Stoller. *Inside the Mafia of Pharma Pricing. www.thebignewsletter.com.* 2024. URL: https://www.thebignewsletter.com/p/inside-the-mafia-of-pharma-pricing.

[92] WMP. *What You Need to Know About the Eyewear and Glasses Monopoly.* 2024. URL: https://www.wmpeyewear.com/blogs/blog/what-you-need-to-know-about-the-eyewear-and-glasses-monopoly?srsltid=AfmBOoqLbl8TawoCo8PLXmRhRJ8AfQmIPouMRS_MlUJ3xB6VAWb_ifPJ.

[93] Frini Btzeni. *Advertising history: exploit and overconsume.* 2023. URL: https://exposure.org.uk/young%5C_peoples%5C_work/history-of-advertising-exploitation-and-overconsumption/.

[94] NOAA Fisheries. *Sustainable Fisheries.* 2019. URL: https://www.fisheries.noaa.gov/topic/sustainable-fisheries.Noaa.gov.

[95] UNEP. *Inside the global effort to save the world's forests.* 2021. URL: https://www.unep.org/news-and-stories/story/inside-global-effort-save-worlds-forests.

[96] FAO. *Sustainable Food and Agriculture. Food and Agriculture Organization of the United Nations*. 2023. URL: https://www.fao.org/sustainability/en/.

[97] UN Environment Programme. *Design for Sustainability: A Step-by-Step Approach*. 2009. URL: https://www.unep.org/resources/report/design-sustainability-step-step-approach.

[98] S. King. *Recycling our way to sustainability*. https://www.nature.com/articles/d41586-022-03646-z. 2022.

[99] Seoul National Univerity. *Program in Environmental Materials Science*. URL: https://cals.snu.ac.kr/en/academics/undergraduate?deptidx=27%5C&majidx=5.

[100] E. Caniglia. *The Path to a Regenerative Future: The Importance of Local Networks and Bioregional Contexts*. 2018. URL: https://www.resilience.org/stories/2018-05-14/the-path-to-a-regenerative-future-the-importance-of-local-networks-and-bioregional-contexts/.

[101] J. et al. Bundy. *The economic potential of metalliferous sub-volcanic brines*. 2021. URL: https://royalsocietypublishing.org/doi/10.1098/rsos.202192.

[102] R. DiResta. *Invisible Rulers*. New York, N.Y.: Public Affairs, 2024.

[103] World Wildlife Fund. *What is the sixth mass extinction and what can we do about it?* 2022. URL: https://www.worldwildlife.org/stories/what-is-the-sixth-mass-extinction-and-what-can-we-do-about-it.

[104] J. Fernando. *Gross Domestic Product (GDP) Formula and How to Use It*. 2025. URL: https://www.investopedia.com/terms/g/gdp.asp.

[105] *Devastating Collapse of Groundfish Fishery Forces a More Sustainable Future | NOAA Fisheries*. 2019. URL: https://www.fisheries.noaa.gov/feature-story/devastating-collapse-groundfish-fishery-forces-more-sustainable-future.

[106] V. Saini. *Drought's Expanding Grip: How 75% of the World Faces a Waterless Future*. 2024. URL: https://climatefactchecks.org/droughts-expanding-grip-how-75-of-the-world-faces-a-waterless-future/.

[107] C. Clinic. *5 Ways Forever Chemicals (PFAS) May Affect Your Health*. 2023. URL: https://health.clevelandclinic.org/what-are-forever-chemicals-pfas.

[108] M. Kozlov. *Landmark study links microplastics to serious health problems*. 2024. URL: https://www.nature.com/articles/d41586-024-00650-3.

[109] A. Kapoor and B. Debroy. *GDP Is Not a Measure of Human Well-Being*. 2019. URL: https://hbr.org/2019/10/gdp-is-not-a-measure-of-human-well-being.

[110] World Health Organization. *Mental health of adolescents*. 2024. URL: https://www.who.int/news-room/fact-sheets/detail/adolescent-mental-health.

[111] Timothy Taylor. *Including illegal Activities in GDP, Drugs, Prostitution, Gambling*. 2021. URL: https://conversableeconomist.com/2021/02/23/.

[112] GNH and Centre Bhutan. *HISTORY OF GNH – GNH Centre Bhutan*. 2024. URL: https://www.gnhcentrebhutan.org/history-of-gnh/.

[113] United Nations Department of Economics and Social Affairs. *Gross National Happiness Index | Department of Economic and Social Affairs*. URL: https://sdgs.un.org/partnerships/gross-national-happiness-index.

[114] *About*. URL: https://gnhusa.org/about/.

[115] World Happiness Report. *About Us*. URL: https://www.worldhappiness.report/about/.

[116] Gallup. *World Happiness Report*. 2024. URL: https://www.gallup.com/analytics/349487/world-happiness-report.aspx.

[117] Christian Krekel and George MacKerron. *How Environmental Quality Affects Our Happiness*. 2020. URL: https://www.worldhappiness.report/ed/2020/how-environmental-quality-affects-our-happiness/.

[118] V Smil. *Harvesting the Biosphere: What We Have Taken from Nature*. Cambridge, MA: MIT Press, 2015.

[119] S. Lee. *Trump-era EPA Meddled in PFAS Science, Agency Watchdog Finds*. 2024. URL: https://news.bloomberglaw.com/environment-and-energy/trump-era-epa-meddled-in-pfas-science-agency-watchdog-finds.

[120] H. Hennerrich. *Here's What Makes Money, Money*. 2021. URL: https://www.stlouisfed.org/open-vault/2021/december/what-is-money.

[121] H. E. Daly, J. B. Cobb, and C. W. Cobb. *For the common good : redirecting the economy toward community, the environment, and a sustainable future*. Boston, MA, 1994.

[122] lyrics.com. *Liza Minnelli - Money, Money Lyrics*. 2025. URL: https://www.lyrics.com/lyrics/Money%5C%20Makes%5C%20the%5C%20World%5C%20Go%5C%20Round.

[123] S. Ross. *Understanding How the Federal Reserve Creates Money*. 2020. URL: https://www.investopedia.com/articles/investing/081415/understanding-how-federal-reserve-creates-money.asp.

[124] Battersby B. Ture E. Larn R. *Tracking the 9 Trillion Global Fiscal Support to Fight COVID-19*. 2020. URL: https://www.imf.org/en/Blogs/Articles/2020/05/20/tracking-the-9-trillion-global-fiscal-support-to-fight-covid-19.

[125] A. Jacobo. *Record number of climate lawsuits target governments and corporations for misleading claims and environmental failures | NationofChange*. 2024. URL: https://www.nationofchange.org/2024/07/01/record-number-of-climate-lawsuits-target-governments-and-corporations-for-misleading-claims-and-environmental-failures/.

[126] L. Paddison. *United Nations report heavily critical of UK's 'increasingly severe crackdowns' on environmental protesters*. 2024. URL: https://www.cnn.com/2024/01/23/climate/environmental-defenders-crackdown-united-nations-uk-climate-intl/index.

[127] Dave Ramsey. *Dave Ramsey — facebook.com*. 2009. URL: https://www.facebook.com/daveramsey/posts/earning-a-lot-of-money-is-not-the-key-to-prosperity-how-you-handle-it-is-dave-ra/139482502686/.

[128] Anon. *Alanis Obomsawin â€"- Wikiquote*. 2023. URL: https://en.wikiquote.org/wiki/Alanis_Obomsawin.

[129] B. Martín-López. *Plural valuation of nature matters for environmental sustainability and justice | Royal Society. royalsociety.org*. 2024. URL: https://royalsociety.org/news-resources/projects/biodiversity/plural-valuation-of-nature-matters-for-environmental-sustainability-and-justice/.

[130] E. Gies. *We can feel our ancestors*. 2024. URL: https://www.theguardian.com/world/2024/sep/23/canada-first-nation-forest-wildfire-logging.

[131] United Nations SEEA. *Ecosystem Accounting | System of Environmental Economic Accounting*. 2012. URL: https://seea.un.org/ecosystem-accounting/.

[132] I. Shatz. *Ingroups and Outgroups: How Social Identity Influences People – Effectiviology*. 2024. URL: https://effectiviology.com/ingroup-outgroup/.

[133] J. Robbins. *Road Hazard: Evidence Mounts on Toxic Pollution from Tires — e360.yale.edu*. 2023. URL: https://e360.yale.edu/features/tire-pollution-toxic-chemicals.

[134] V. Chub. *How Many Trucks Are There In The World? (2025) - Vehicle Help*. 2025. URL: https://vehiclehelp.com/how-many-trucks-are-there-in-the-world/.

[135] T. M. Lenton et al. *Climate tipping points — too risky to bet against*. 2019. URL: https://www.nature.com/articles/d41586-019-03595-0.

[136] David I Armstong and Sina Lariani. *1: Earth System Tipping Points*. 2023. URL: https://report-2023.global-tipping-points.org/section1/1-earth-system-tipping-points/.

[137] Anon. *Fossil Fuel Subsidies Surged to Record $7 Trillion*. URL: https://www.imf.org/en/Blogs/Articles/2023/08/24/fossil-fuel-subsidies-surged-to-record-7-trillion.

[138] So Kat. *Climate Deniers of the 118 Congress*. 2024. URL: https://www.americanprogress.org/article/climate-deniers-of-the-118th-congress/.

[139] F. Harvey. *'Massive disinformation campaign' is slowing global transition to green energy. the Guardian*. 2024. URL: https://www.theguardian.com/environment/article/2024/aug/08/fossil-fuel-industry-using-disinformation-campaign-to-slow-green-transition-says-un.

[140] Jesse Hirsh. *195: The Great Trump Depression and the End of Money*. 2025. URL: https://metaviews.substack.com/p/195-the-great-trump-depression-and.

[141] United Nations. *The 17 Sustainable Development Goals*. 2024. URL: https://sdgs.un.org/goals.

[142] dashboards.sdgindex.org. *Sustainable Development Report 2021*. 2023. URL: https://dashboards.sdgindex.org/chapters.

[143] Sustainable Development Report. *Rankings — The overall performance of all 193 UN Member States*. 2024. URL: https://dashboards.sdgindex.org/rankings.

[144] N. Smith. *neoliberalism — polical and social science*. 2023. URL: https://www.britannica.com/money/neoliberalism.

[145] Clea Simon. *New clues about how and why the Maya culture collapsed*. 2020. URL: https://news.harvard.edu/gazette/story/2020/02/new-clues-about-how-and-why-the-maya-culture-collapsed/.

[146] A. Kravtsova. *'It's like I drew a door and disappeared through it' M*. 2021. URL: https://meduza.io/en/feature/2021/04/01/it-s-like-i-drew-a-door-and-disappeared-through-it.

[147] National Low Income Housing Coalition. *Out of Reach*. 2024. URL: https://nlihc.org/oor/about.

[148] C.-T. Kuo and I. Kawachi. *County-Level Income Inequality, Social Mobility, and Deaths of Despair in the US, 2000-2019*. 2023. URL: https://jamanetwork.com/journals/jamanetworkopen/fullarticle/2807161.

[149] M. Farrell and A. Das. *Pay for Lawyers Is So High People Are Comparing It to the N*. 2024. URL: https://www.nytimes.com/2024/07/01/business/law-firm-pay-salary.html.

[150] B. Ballou. *Plunder*. New York, NY: PublicAffairs, 2023.

[151] R. Prete. *How private equity firms make money | PitchBook*. 2021. URL: https://pitchbook.com/news/articles/how-private-equity-firms-make-money.

[152] B. Ballou. *Private Equity Is Gutting America — and Getting Away With It*. 2023. URL: https://www.nytimes.com/2023/04/28/opinion/private-equity.html.

[153] Thomas D. and Sabater A. *US private equity portfolio company bankruptcies spiked to record high in 2023*. 2024. URL: https://www.spglobal.com/marketintelligence/en/news-insights/latest-news-headlines/us-private-equity-portfolio-company-bankruptcies-spiked-to-record-high-in-2023-80000182.

[154] David Gelles, Lisa Friedman, and Brad Plumer. *How Trump Has Undermined U.S. Climate Policy*. 2025. URL: https://www.nytimes.com/2025/03/02/climate/trump-us-climate-policy-changes.html.

[155] Adam B. Smith. *2024: An active year of U.S. billion-dollar weather and climate disasters*. 2025. URL: https://www.climate.gov/news-features/blogs/beyond-data/2024-active-year-us-billion-dollar-weather-and-climate-disasters.

[156] E. Biss. *The Theft of the Commons*. 2022. URL: https://www.newyorker.com/culture/essay/the-theft-of-the-commons.

[157] History on the Net. *Industrial Revolution Working Conditions - History*. 2025. URL: https://www.historyonthenet.com/industrial-revolution-working-conditions.

[158] anti-slavery. *What Is Modern Slavery?* 2024. URL: https://www.antislavery.org/slavery-today/modern-slavery/.

[159] Crownhart C. *Last summer was the hottest in 2,000 years*. 2024. URL: https://www.technologyreview.com/2024/05/16/1092507/tree-rings-climate-data/.

[160] B. Girod. *Heat-related deaths in Florida have increased 88% in 3 years as temps continue to rise*. 2023. URL: https://www.pnj.com/story/news/2023/08/04/heat-related-deaths-in-florida-have-increased-88-in-three-years/70519895007/.

[161] Public Citizen. *DeSantis Signs Law Banning Water Breaks and Cooling Measures from Florida Workers*. 2024. URL: https://www.citizen.org/news/desantis-signs-law-banning-water-breaks-and-cooling-measures-from-florida-workers/.

[162] B. Girod. *Heat-related deaths in Florida have increased 88% in 3 years as temps continue to rise*. 2023. URL: https://www.pnj.com/story/news/2023/08/04/heat-related-deaths-in-florida-have-increased-88-in-three-years/70519895007/.

[163] Borenstein M. Wildeman K. Snow A. *AP analysis finds 2023 set record for US heat deaths, killing in areas that used to handle the heat*. 2024. URL: https://apnews.com/article/record-heat-deadly-climate-change-humidity-south-11de21a526e1cbe7e306c47c2f12438d.

[164] T. Dohmke. *100 million developers and counting*. 2023. URL: https://github.blog/2023-01-25-100-million-developers-and-counting/.

[165] J. Sommer. *A New Measure Shows CEO Pay at Even More Astronomical Levels*. 2024. URL: https://www.nytimes.com/2024/06/07/business/ceo-pay-compensation-stock-market.html.

[166] Koenig D. and Richer A.D. *Here's what to know about Boeing agreeing to plead guilty to fraud in 737 Max crashes*. 2024. URL: https://apnews.com/article/boeing-guilty-plea-fraud-justice-victims-fd309f4cc643637ebd9637fd5864dbb9.

[167] T. Rosenbluth and C. Jewett. *Boar's Head Plant Tied to 9 Deaths Had Mold, Leaky Pipes and Flies*. 2024. URL: https://www.nytimes.com/2024/08/30/health/listeria-outbreak-boars-head-deli-meat.html.

[168] A. Case and A. Deaton. *Deaths of Despair and the Future of Capitalism*. Princeton, NJ: Princeton University Press, 2020.

[169] Oxfam. *What is global inequality?* 2022. URL: https://www.oxfamamerica.org/explore/stories/what-is-global-inequality/.

[170] Oxfam. *What is global inequality? www.oxfamamerica.org*. 2022. URL: https://www.oxfamamerica.org/explore/stories/what-is-global-inequality/.

[171] IEA. *Executive summary – World Energy Employment 2023 – Analysis*. 2023. URL: https://www.iea.org/reports/world-energy-employment-2023/executive-summary.

[172] International Energy Agency (IEA). *World Energy Employment 2023 — Executive summary*. 2023. URL: https://www.iea.org/reports/world-energy-employment-2023/executive-summary.

[173] M. Jones et al. *Engineered Mycelium Composite Construction Materials from Fungal biorefineries: a Critical Review*. 2020. URL: https://www.sciencedirect.com/science/article/pii/S0264127519308354?via%5C%3Dihub.

[174] worldometer. *GDP by Country*. 2023. URL: https://www.worldometers.info/gdp/gdp-by-country/.

[175] Center for Child Care Employment. *Early Educator Pay & Economic Insecurity Across the States*. 2024. URL: https://cscce.berkeley.edu/workforce-index-2024/the-early-childhood-educator-workforce/early-educator-pay-economic-insecurity-across-the-states/.

[176] Keith-Jennings B. and V. Palacios. *SNAP Helps Millions of Low-Wage Workers*. 2017. URL: https://www.cbpp.org/research/snap-helps-millions-of-low-wage-workers.

[177] M. Farrell and A. Das. *Pay for Lawyers Is So High People Are Comparing It to the N*. 2024. URL: https://www.nytimes.com/2024/07/01/business/law-firm-pay-salary.html.

[178] D. Graeber. *Bullshit Jobs: A Theory*. New York, NY: Simon & Schuster, 2018.

[179] Center for Guaranteed Income Research, Stanford Basic Income Lab, and Mayors for a Guaranteed Income. *The Guaranteed Income Pilots Dashboard*. 2022. URL: https://guaranteedincome.us/.

[180] K. Amadeo. *Living Wage: How Much Do You Need to Live in America?* 2022. URL: https://www.thebalancemoney.com/living-wage-3305771.

[181] D. Broom. *Four-day work week trial in Spain leads to healthier workers, less pollution*. 2023. URL: https://www.weforum.org/agenda/2023/10/surprising-benefits-four-day-week/

[182] E. Griffin. *Every company should be owned by its employees. www.elysian.press*. 2024. URL: https://www.elysian.press/p/employee-ownership.

[183] Democracy at Work Institute. *What is a Worker Cooperative?* 2024. URL: https://institute.coop/what-worker-cooperative-0.

[184] Z. Abrams. *The Urgent Necessity for Paid Parental Leave*. 2022. URL: https://www.apa.org/monitor/2022/04/feature-parental-leave.

[185] E. Ortiz-Ospina and M. Roser. *Child Labor. Our World in Data*. 2016. URL: https://ourworldindata.org/child-labor.

[186] *Community-baed Natural Resource Management (CBNRM)*. URL: https://biodiversitylinks.org/library/resources/rmp/collections/folder-for-the-environment-collections/community-based-natural-resource-management-cbnrm/.

[187] Roosevelt Institute. *Why Workers On Corporate Boards Just Makes Sense*. 2018. URL: https://rooseveltinstitute.org/2018/08/14/why-workers-on-corporate-boards-just-makes-sense/.

[188] Naidu S. and Dube A. *Monopsony Power in Labor Markets*. 2024. URL: https://www.nber.org/reporter/2024number1/monopsony-power-labor-markets.

[189] Rajkarnikar P.J. *Mad Money: The Financialization and Rising Inequality of the US Economy Economics in Context Initiative*. 2022. URL: https://www.bu.edu/eci/2022/08/02/mad-money-the-financialization-and-rising-inequality-of-the-us-economy/.

[190] Institute for Policy Studies. *Inequality and Taxes*. 2024. URL: https://inequality.org/facts/taxes-inequality-in-united-states/.

[191] Aaron. *To reduce inequality, tax inheritances*. 2019. URL: https://www.brookings.edu/articles/to-reduce-inequality-tax-inheritances/.

[192] De Witte M. *This is how we can stamp out supply-chain slavery*. 2022. URL: https://www.weforum.org/agenda/2022/01/stanford-global-supply-chains-modern-slavery/.

[193] T. Morlà-Folch et al. *The Mondragon Case: Companies Addressing Social Impact and Dialogic Methodologies*. 2021. URL: https://journals.sagepub.com/doi/10.1177/16094069211058614.

[194] Shift. *The Case for Living Wages How Paying Living Wages Improves Business Performance and Tackles Poverty*. 2022. URL: https://shiftproject.org/resource/case-for-living-wages/.

[195] U. Rana. *A long weekend every week? Why Japan is the latest urging firms to consider it*. 2024. URL: https://globalnews.ca/news/10730980/japan-four-day-work-week/.

[196] Charles Hugh Smith. *There's Just One Problem: AI Isn't Intelligent, and That's a Systemic Risk*. 2024. URL: http://charleshughsmith.blogspot.com/2024/08/theres-just-one-problem-ai-isnt.html.

[197] A. Narayanan and S. Kapoor. *AI existential risk probabilities are too unreliable to inform policy*. 2024. URL: https://www.aisnakeoil.com/p/ai-existential-risk-probabilities.%20Aisnakeoil.com.

[198] Alex Wallerstein. *Counting the dead at Hiroshima and Nagasaki*. 2020. URL: https://thebulletin.org/2020/08/counting-the-dead-at-hiroshima-and-nagasaki/.

[199] Lori Esposito Murry. *The Risk of Nuclear War Continues to Rise*. 2025. URL: https://www.cfr.org/article/risk-nuclear-war-continues-rise.

[200] Jennie Cohen. *History's 6 Worst Nuclear Disasters*. 2025. URL: https://www.history.com/articles/historys-worst-nuclear-disasters.

[201] Davi Ottenheimer. *MCP as Nostalgia Misses Forest for Trees: The Danger Inside Anil Dash's Celebrations*. 2025. URL: https://www.flyingpenguin.com/?p=70421.

[202] Priyadarshini Ravichandran Ravichandran. *How WhatsApp Leads Mobs to Murder in India*. 2017. URL: https://www.nytimes.com/interactive/2018/07/18/technology/whatsapp-india-killingss.html.

[203] Kikkidan Yibeltal and Wycliffe Mula. *Facebook's algorithms 'supercharged' hate speech in Ethiopia's Tigray conflict*. 2023. URL: https://www.bbc.com/news/world-africa-67275219.

[204] Megan Squire. *Digital Threat Report: Mass Shooter Videos*. 2024. URL: https://www.splcenter.org/resources/hatewatch/digital-threat-report-mass-shooter-videos/.

[205] Amnesty International. *How TikTok encourages self-harm & suicidal ideation*. 2025. URL: https://edri.org/our-work/tiktoks-for-you-feed-risks-pushing-children-and-young-people-towards-harmful-mental-health-content/.

[206] David Ingram. *Verified pro-Nazi X accounts flourish under Elon Musk*. 2024. URL: https://www.nbcnews.com/tech/social-media/x-twitter-elon-musk-nazi-extremist-white-nationalist-accounts-rcna145020.

[207] B. Sopelsa and J. Yurcaba. *LGBTQ youth suicide prevention group leaves X after uptick in 'hate & vitriol'*. 2023. URL: https://www.nbcnews.com/nbc-out/out-news/lgbtq-youth-suicide-prevention-group-leaves-x-uptick-hate-vitriol-rcna124501.

[208] United States Attorney's Office Southern District of Indiana. *FBI and Partners Issue National Public Safety Alert on Sextortion Schemes.* 2023. URL: https://www.justice.gov/usao-sdin/pr/fbi-and-partners-issue-national-public-safety-alert-sextortion-schemes#:~:text=Over%5C%20the%5C%20past%5C%20year%5C%2C%5C%20law,to%5C%20have%5C%20died%5C%20by%5C%20suicide..

[209] Goldman Sacks. *Will the $1 trillion of generative AI investment pay off?* 2024. URL: https://www.goldmansachs.com/insights/articles/will-the-1-trillion-of-generative-ai-investment-pay-off.

[210] Stephen Rivers. *Tesla On FSD Suddenly Swerves And Crashes Into A Tree.* 2025. URL: https://www.carscoops.com/2025/05/tesla-fsd-crash-video-swerve-tree/.

[211] Renee Barnes and et al. Riedel Aimee. *Disinformation and deepfakes played a part in the Uelection. Australia should expect the same.* 2024. URL: https://theconversation.com/disinformation-and-deepfakes-played-a-part-in-the-us-election-australia-should-expect-the-same-243373.

[212] Lydia Saad. *More U.S. Workers Fear Technology Making Their Jobs Obsolete.* 20232023. URL: https://news.gallup.com/poll/510551/workers-fear-technology-making-jobs-obsolete.aspx.

[213] Goldman Sachs. *AI to drive 165% increase in data center power demand by 2030.* 2025. URL: https://www.goldmansachs.com/insights/articles/ai-to-drive-165-increase-in-data-center-power-demand-by-2030.

[214] Daniel Howley. *Big Tech is going all in on nuclear power as sustainability concerns around AI grow.* 2024. URL: https://finance.yahoo.com/news/big-tech-is-going-all-in-on-nuclear-power-as-sustainability-concerns-around-ai-grow-201418764.html.

[215] World Nuclear Association. *Plans For New Reactors Worldwide.* 2025. URL: https://world-nuclear.org/information-library/current-and-future-generation/plans-for-new-reactors-worldwide.

[216] GeeksforGeeks. *Narrow AI vs General AI.* 2024. URL: https://www.geeksforgeeks.org/artificial-intelligence/narrow-ai-vs-general-ai/.

[217] Stephen Fry. *AI: A Means to an End or a Means to Our End?* 2024. URL: https://stephenfry.substack.com/p/ai-a-means-to-an-end-or-a-means-to.

[218] Chariot Energy. *How Grid Scale Energy Storage is Revolutionizing Power Management.* 2024. URL: https://chariotenergy.com/blog/maximizing-renewable-energy-with-grid-scale-storage-solutions/.

[219] European Environment Agency. *Electric vehicles*. 2024. URL: https://www.eea.europa.eu/en/topics/in-depth/electric-vehicles.

[220] Qiong Shen, Yuxi Pan, and Yanchao Feng. *The impacts of high-speed railway on environmental sustainability: quasi-experimental evidence from China*. 2023. URL: https://www.nature.com/articles/s41599-023-02135-6.

[221] V Thamarai et al. *Carbon mitigation in agriculture: Pioneering technologies for a sustainable food system*. 2024. URL: https://www.sciencedirect.com/science/article/abs/pii/S0924224424001535.

[222] Soil Association. *What is organic?* URL: https://www.soilassociation.org/take-action/organic-living/what-is-organic/.

[223] Bernt Nelson. *The Evolution of Precision Agriculture and Policy Implications*. 2023. URL: https://www.fb.org/market-intel/the-evolution-of-precision-agriculture-and-policy-implications.

[224] Bryson Funk. *Vertical Farming: Everything You Need to Know*. 2023. URL: https://www.edengreen.com/blog-collection/what-is-vertical-farming.

[225] Rodale Institue. *Carbon Sequestration*. 2025. URL: https://rodaleinstitute.org/why-organic/issues-and-priorities/carbon-sequestration/.

[226] Tharika Lecamwasam. *Three emerging technologies for low-carbon concrete*. 2024. URL: https://www.climateworks.org/blog/three-emerging-technologies-for-low-carbon-concrete/.

[227] Watsom. Natasha. *ConcreteZero: providing a solid foundation for low carbon construction*. 2023. URL: https://www.theclimategroup.org/our-work/news/concretezero-providing-solid-foundation-low-carbon-construction.

[228] Jeffrey Kluger. *Companies Knew the Dangers of PFAS 'Forever Chemicals'—and Kept Them Secret*. 2023. URL: https://time.com/6284266/pfas-forever-chemicals-manufacturers-kept-secret/.

[229] V. Tangermann. *Scientists Create Robot Controlled by Blob of Human Brain Cells*. 2024. URL: https://futurism.com/neoscope/robot-human-brain-cells.

[230] Daan Walter, Sam Butler-Sloss, and Kinsmill Bond. *Rewiring the energy debate*. 2025. URL: https://electrotechrevolution.substack.com/p/rewiring-the-energy-debate.

[231] S.N. Botman. *Revisiting the broadcast public interest standard in communications law and regulation*. 2017. URL: https://www.brookings.edu/articles/revisiting-the-broadcast-public-interest-standard-in-communications-law-and-regulation/.

[232] LegalClarity Team. *What Is the Fairness Doctrine and What Does It Mean for Broadcasterrs*. 2025. URL: https://legalclarity.org/what-is-the-fairness-doctrine-and-what%20-does-it-mean-for-broadcasters/.

[233] S. Rendall. *The Fairness Doctrine: How We Lost It and Why We Need It Back*. 2025. URL: https://sisyphuslitmag.org/2018/07/the-fairness-doctrine-how-we-lost-it-and-why-we-need-it-back/.

[234] D. Thussu. *News as Entertainment: The Rise of Global Infotainment*. 2007. URL: https://sk.sagepub.com/books/news-as-entertainment.

[235] B. L. Fossen. *The science of product placements – and why some work better than others*. 2021. URL: https://theconversation.com/the-science-of-product-placements-and-why-some-work-better-than-others-165435.

[236] Ani Petrosyan. *Worldwide digital population 2025*. 2025. URL: https://www.statista.com/statistics/617136/digital-population-worldwide/.

[237] *Forrester: Cybercrime to cost $12 trillion in 2025*. 2024. URL: https://dig.watch/updates/forrester-cybercrime-to-cost-12-trillion-in-2025.

[238] Adrian Shabaz and Allie Funk. *Social Media Surveillance*. 2025. URL: https://freedomhouse.org/report/freedom-on-the-net/2019/the-crisis-of-social-media/social-media-surveillance.

[239] R. DiResta. *Invisible Rulers*. New York, NY: PublicAffairs, 2024.

[240] Mike Isaac and Theodore Shleifer. *Meta Says It Will End Its Fact-Checking Program on Social Media Posts*. 2025. URL: https://www.nytimes.com/live/2025/01/07/business/meta-fact-checking.

[241] E. Sullivan. *Election Workers Face Flood of Threats, but Charges Are Few*. 2024. URL: https://www.nytimes.com/2024/04/13/us/politics/election-workers-threats.html.

[242] *Internet Crime Complaint Center (IC3) | Cyber Criminals Impersonating Brands Using Search Engine Advertisement Services to Defraud Users*. 2022. URL: https://www.ic3.gov/Media/Y2022/PSA221221.

[243] M. Brenan. *Americans' Trust in Mass Media Edges Down to 41*. 2024. URL: https://news.gallup.com/poll/651977/americans-trust-media-remains-trend-low.aspx.

[244] Walsh C. *Truth vs. 'truthiness'*. 2015. URL: https://news.harvard.edu/gazette/story/2015/01/truth-vs-truthiness/.

[245] T. Mak and Hay A. *Are we in Cold War 2.0? www.counteroffensive.news.* 2024. URL: https://www.counteroffensive.news/p/are-we-in-cold-war-20.

[246] *How to Stop Disinformation.* 2022. URL: https://www.ucsusa.org/resources/how-stop-disinformation.

[247] *United Nations Global Principles for Information Integrity.* URL: https://www.un.org/en/information-integrity/global-principles.

[248] Global Financial Integrity. *Shadow Figures: The data of transnational crime.* 2025. URL: https://gfintegrity.org/wp-content/uploads/2025/10/Shadow-Figures-Final.pdf.

[249] Global Witness. *More than 2,100 land and environmental defenders killed globally between 2012 and 2023.* 2024. URL: https://globalwitness.org/en/press-releases/more-than-2100-land-and-environmental-defenders-killed-globally-between-2012-and-2023/.

[250] Carwright M.M. *The Suppression of Individual Liberty | Psychology Today.* 2016. URL: https://www.psychologytoday.com/us/blog/food-thought/201605/the-suppression-individual-liberty.

[251] Khan E. *Top 15 Most Cruel Rulers Ever in History - Wonderslist.* 2023. URL: https://www.wonderslist.com/10-most-cruel-rulers-ever-in-history/.

[252] Sam Haselby. *Governing for the planet.* 2024. URL: https://aeon.co/essays/why-planetary-problems-need-a-new-approach-to-politics.

[253] Anon. *Task Force for a Global Alliance against Hunger and Poverty.* 2024. URL: https://www.wfp.org/news/wfp-joins-global-alliance-against-hunger-and-poverty-founding-member.

[254] World Food Program. *WFP joins the Global Alliance Against Hunger and Poverty as founding member.* 2024. URL: https://www.wfp.org/news/wfp-joins-global-alliance-against-hunger-and-poverty-founding-member.

[255] G. Brown. *It's August 2024 – and our world is at a turning point. Here's what we should do now.* 2024. URL: https://www.theguardian.com/commentisfree/article/2024/aug/24/august-2024 world turning point-political-environmental-threats-urgent.

[256] W. Morris. *News from Nowhere.* 2024. URL: https://standardebooks.org/ebooks/william-morris/news-from-nowhere..

[257] R. E. Blanton. *Political collapse: Lessons from fallen empires | NationofChange.* 2024. URL: https://www.nationofchange.org/2024/09/04/political-collapse-lessons-from-fallen-empires/.%20Nationofchange.org.

[258] G. Dewar. *Babies expect fairness, and prefer people who behave with fairness.* 2020. URL: https://parentingscience.com/babies-expect-fairness/.

[259] *Scientists and climate change: Extreme concern and high level of engagement.* 2024. URL: https://phys.org/news/2024-08-scientists-climate-extreme-high-engagement.html.

[260] University of Cambridge. *Kate Raworth – Cambridge Institute for Sustainability Leadership.* 2025. URL: https://www.cisl.cam.ac.uk/directory/kate-raworth.

[261] *Herman Daly.* 2019. URL: https://en.wikipedia.org/wiki/Herman%5C_Daly.

[262] *Samuel Alexander.* 2018. URL: https://simplicityinstitute.org/samuel-alexander.

[263] *Lorenzo Fioramonti.* 2024. URL: https://en.wikipedia.org/wiki/Lorenzo%5C_Fioramonti.

[264] *Beth Stratford.* 2024. URL: https://neweconomics.org/profile/beth-stratford.

[265] NobelPrize.org. *The Sveriges Riksbank Prize in Economic Sciences in Memory of Alfred Nobel 1971.* 2019. URL: https://www.nobelprize.org/prizes/economic-sciences/1971/kuznets/facts/.

[266] Estrellita García. *Manifesto for a post-growth economy - Economy for the common good. Economy for the common good.* 2023. URL: https://www.econgood.org/manifesto-for-a-post-growth-economy/.

[267] J. Mecklin. *A Time of Unprecedented Danger: It Is 90 Seconds to Midnight. Bulletin of the Atomic Scientists.* 2024. URL: https://thebulletin.org/doomsday-clock/current-time/.

[268] *A quote by William Thomson.* URL: https://www.goodreads.com/quotes/166961-when-you-can-measure-what-you-are-speaking-about-and.

[269] D. Marchese. *This Pioneering Economist Says Our Obsession With Growth Must End.* 2022. URL: https://www.nytimes.com/interactive/2022/07/18/magazine/herman-daly-interview.html.

[270] C. Anthony. *Which 10 Countries Have the Most Natural Resources? Investopedia.* 2022. URL: https://www.investopedia.com/articles/markets-economy/090516/10-countries-most-natural-resources.asp.

[271] Gregory S.Y. *The mental health benefits of nature: Spending time outdoors to refresh your mind.* 2024. URL: https://mcpress.mayoclinic.org/mental-health/the-mental-health-benefits-of-nature-spending-time-outdoors-to-refresh-your-mind/.

[272] Shankar Sri Ravi. *Our Spiritual Connection to Nature.* 2020. URL: https://wisdom.srisriravishankar.org/our-spiritual-connection-to-nature/.

[273] E. Spencer. *Plankton: Small Organisms with a Big Role in the Ocean - Ocean Conservancy.* 2019. URL: https://oceanconservancy.org/blog/2019/08/09/plankton-small-organism-big-role/.

[274] M. Ruta and A. J. Venables. *International Trade in Natural Resources: Practice and Policy.* 2012. URL: https://www.annualreviews.org/content/journals/10.1146/annurev-resource-110811-114526.

[275] Ted O'Callahan. *The Decades-Long Effort to Account for the Value of Natural Resources.* 2023. URL: https://insights.som.yale.edu/insights/the-decades-long-effort-to-account-for-the-value-of-natural-resources.

[276] David L. *Moon mining gains momentum as private companies plan for a lunar economy.* 2023. URL: https://www.space.com/moon-mining-gains-momentum.%20D.%20Moon%20mining%20gains%20momentum%20as%20private%20companies%20plan%20for%20a%20lunar%20economy.%20Space.com.

[277] H. E. Daly. *For The Common Good by Herman E.* 2024. URL: https://www.penguinrandomhouse.com/books/203945/for-the-common-good-by-beacon/.

[278] Fenichel E. *The Decades-Long Effort to Account for the Value of Natural Resources.* 2023. URL: https://insights.som.yale.edu/insights/the-decades-long-effort-to-account-for-the-value-of-natural-resources.

[279] L. Landrigan. *How the Codfish Started the American Revolution.* 2024. URL: https://newenglandhistoricalsociety.com/how-codfish-started-the-american-revolution/.

[280] Wikipedia. *Collapse of the Atlantic northwest cod fishery.* 2024. URL: https://en.wikipedia.org/wiki/Collapse_of_the_Atlantic_northwest_cod_fishery.

[281] Margaret Alic. *Georges Bank (Collapse of the Ground Fishery).* 2012. URL: https://www.encyclopedia.com/environment/encyclopedias-almanacs-transcripts-and-maps/georges-bank-collapse-ground-fishery.

[282] Saving Seafood. *40 years of change: For fishing industry, the spring of 1976 was the start of a new era*. 2025. URL: https://www.savingseafood.org/other-news/40-years-change-fishing-industry-spring-1976-start-new-era/.

[283] Chris Chase. *Canada reopens Northern cod fishery after 32-year moratorium*. 2024. URL: https://www.seafoodsource.com/news/supply-trade/canada-reopens-northern-cod-fishery-after-32-year-moratorium.

[284] AP News. *Cod fishing limits to be slashed again, regulators say*. 2021. URL: https://apnews.com/article/business-environment-maine-overfishing-5ab09cd517323c4112a9dd4c1fcbf601.

[285] The Maine Climate Council. *Climate Impacts*. URL: https://www.maine.gov/climateplan/climate-impacts.

[286] D. Anderson. *New Plan Could Rebuild GOM Cod Stocks By 2033 - The Fisherman*. 2023. URL: https://www.thefisherman.com/article/new-plan-could-rebuild-gom-cod-stocks-by-2033/.

[287] *Cod Research in Massachusett and the Release of a New StoryMap Mass.gov*. 2022. URL: https://www.mass.gov/news/cod-research-in-massachusetts-and-the-release-of-a-new-storymap.

[288] *Natural resources - The World Factbook*. 2024. URL: https://www.cia.gov/the-world-factbook/field/natural-resources/.

[289] *The World Bank Commodities Price Data – Pink Sheet*. 2024. URL: https://thedocs.worldbank.org/en/doc/5d903e848db1d1b83e0ec8f744e55570-0350012021/related/CMO-Pink-Sheet-August-2024.pdf?%5C_gl=1.

[290] *Countries With the Most Natural Resources*. 2024. URL: https://www.worldatlas.com/articles/countries-with-the-most-natural-resources.html.

[291] M. L. Ross. *What Have We Learned about the Resource Curse?* https://www.annualreviews.org/content/journals/10.1146/annurev-polisci-052213-040359. 2015.

[292] Bill McKibben. *On Democracy, Liberated Energy, and the Perfect Moment*. 2025. URL: https://billmckibben.substack.com/p/on-democracy-liberated-energy-and.

[293] UN environment programme. *Bend the trend Pathways to a liveable planet as resource use spikes*. 2024. URL: https://wedocs.unep.org/bitstream/handle/20.500.11822/44901/Global-Resource-Outlook%5C_2024.pdf?sequence=5.

[294] P. Mullan. *The myth of a 'green' Karl Marx*. 2024. URL: https://www.spiked-online.com/2024/08/25/the-myth-of-a-green-karl-marx/.

[295] United Nations Statistics Division. *Progress Chart*. 2023. URL: https://unstats.un.org/sdgs/report/2023/progress-chart/.

[296] O. Milman. *It's mindblowing. US meteorologists face death threats as hurricane conspiracies surge*. 2024. URL: https://www.theguardian.com/us-news/2024/oct/11/meteorologists-death-threats-hurricane-conspiracies-misinformation.

[297] William J. Ripple et al. 2024. URL: https://academic.oup.com/bioscience/advance-article/doi/10.1093/biosci/biae087/7808595.

[298] Oliver Milman and Jonathan Watts. *Global heating makes hurricanes like Helene twice as likely, data shows*. 2024. URL: https://www.theguardian.com/environment/2024/oct/08/hurricane-helene-milton-global-heating-study.

[299] B. McKibben. *Our dystopian climate isn't just about fires and floods*. 2024. URL: https://www.theguardian.com/commentisfree/2024/oct/09/climate-crisis-hurricane-election.

[300] Celeste Sangster. *Guardians of the Earth: The essential role of Indigenous Peoples in the protection of natural resources*. 2025. URL: https://globalgoals.org/news/guardians-of-the-earth-the-essential-role-of-indigenous-peoples-in-the-protection-of-natural-resources/.

[301] Ashley Crossman. *Herbert Spencer's Contributions to Sociology*. 2025. URL: https://www.thoughtco.com/herbert-spencer-3026492.

[302] German History in Documents and Images. *Nazi and American Eugenicists (1937)*. 2024. URL: https://germanhistorydocs.org/en/nazi-germany-1933-1945/nazi-and-american-eugenicists-1937.

[303] S.L. Spanoidis. *Poets' Corner - Alfred, Lord Tennyson - In Memoriam*. 2020. URL: https://www.theotherpages.org/poems/books/tennyson/tennyson04.html.

[304] Trees for Life. *Symbiosis in nature*. 2025. URL: https://treesforlife.org.uk/into-the-forest/habitats-and-ecology/ecology/symbiosis/.

[305] Katharine Hayhoe. *Today's floods are tomorrow's warnings*. 2025. URL: https://www.talkingclimate.ca/p/todays-floods-are-tomorrows-warnings.

[306] T. Silver. *Three Worlds, Three Views: Culture and Environmental Change in the Colonial South, Nature Transformed*. 2019. URL: http://nationalhumanitiescenter.org/tserve/nattrans/ntuseland/essays/threeworlds.htm.

[307] A. McQuatters-Gollop. *Small but mighty, plankton are some of the most powerful creatures on Earth*. 2024. URL: https://theconversation.com/small-but-mighty-plankton-are-some-of-the-most-powerful-creatures-on-earth-229244.

[308] J. Parnell. *How plankton helped create the Earth's mountains 2 billion years ago*. 2021. URL: https://theconversation.com/how-plankton-helped-create-the-earths-mountains-2-billion-years-ago-173076.

[309] P. Dockrill. *Scientists Reveal a Massive Biosphere of Life Hidden Under Earth's Surface*. 2018. URL: https://www.sciencealert.com/scientists-lift-lid-on-massive-biosphere-of-life-hidden-under-earth-s-surface.

[310] Baker H. *Discovery of 'hidden world' under Antarctic ice has scientists 'jumping for joy'*. 2022. URL: https://www.livescience.com/hidden-ecosystem-under-antarctic-ice.

[311] Abe S. *NASA Astrobiology*. 2024. URL: https://astrobiology.nasa.gov/news/nasa-research-focuses-on-yellowstones-hot-springs-and-compares-findings-to-rocks-from-mars/.

[312] Abdullahi Tsanni. *More than 100 bacteria species can flourish in microwave ovens — sciencenews.org*. 2024. URL: https://www.sciencenews.org/article/some-bacteria-flourish-microwave-ovens.

[313] David Nutt. *Biohybrid robots controlled by electrical impulses — in mushrooms*. 2024. URL: https://news.cornell.edu/stories/2024/08/biohybrid-robots-controlled-electrical-impulses-mushrooms.

[314] *How many bacteria vs human cells are in the body?* 2016. URL: http://www.microbiomeinstitute.org/blog/2016/1/20/how-many-bacterial-vs-human-cells-are-in-the-body.%20D.%20The%20American%20Microbiome%20Institute.%20The%20American%20Microbiome%20Institute.

[315] K. Raworth. *Doughnut economics: seven ways to think like a 21st-century economist*. New York, N.Y., 2017.

[316] E. Walsh. *Energy Flow (Ecosystem): Definition, Process & Examples*. 2019. URL: https://sciencing.com/energy-flow-ecosystem-definition-process-examples-with-diagram-13719231.html.

[317] R. Chaplin-Kramer et al. *Global modeling of nature's contributions to people*. 2019. URL: https://www.science.org/doi/10.1126/science.aaw3372.

[318] Center for a New Economics. *New Alchemy: Ecological Design 'Mecca'*. 2022. URL: https://centerforneweconomics.org/newsletters/new-alchemy-ecological-design-mecca/.

[319] John Todd Ecological Design. *John Todd Ecological Design*. 2016. URL: https://www.toddecological.com/.

[320] J. Todd. *Healing Earth: An Ecologist's Journey of Innovation and Environmental Stewardship*. 2019. URL: https://www.northatlanticbooks.com/shop/healing-earth/.

[321] R. Schleeter. *The Great Green Wall National Geographic Society. education.nationalgeographic.org*. 2022. URL: https://education.nationalgeographic.org/resource/great-green-wall/.

[322] UN Decade on Restoration. *RESTORE: The Great Green Wall*. 2021. URL: https://www.decadeonrestoration.org/restore-great-green-wall.

[323] UN Decade on Restoration. *About the UN Decade*. URL: https://www.decadeonrestoration.org/about-un-decade.

[324] W. Anderson. *Pressing Questions About Ecosystem Restoration*. 2021. URL: https://www.wri.org/insights/ecosystem-restoration-questions.%20Answered.%20www.wri.org.

[325] Carter Brandon and et al. Kratzer Bradley. *Strengthening the Investment Case for Climate Adaptation: A Triple Dividend Approach*. 2025. URL: https://www.wri.org/research/climate-adaptation-investment-case.

[326] Klaus V. H. and Kiehl K. *A conceptual framework for urban ecological restoration and rehabilitation*. https://www.sciencedirect.com/science/article/pii/S1439179121000372?via%3Dihub. 2021.

[327] UNESCO. *Imminent risk of a global water crisis, warns the UN World Water Development Report 2023*. 2023. URL: https://www.unesco.org/en/articles/imminent-risk-global-water-crisis-warns-un-world-water-development-report-2023.%20Unesco.org.

[328] *Selected Significant Climate Anomalies and Events: 2024*. 2024. URL: https://www.ncei.noaa.gov/access/monitoring/monthly-report/global/202408.

[329] Smithfield. *The world's 4th largest lake has dried up to less than 10% its size and is now a giant desert, dotted with shipwrecks*. 2015. URL: https://www.thevintagenews.com/2015/09/29/the-worlds-4th-largest-lake-has-dried-up-to-less-than-10-its-size-and-is-now-a-giant-desert-dotted-with-shipwrecks/.

[330] Traxler V. *'We used to sail and fish and play': how did an Argentinian lake the size of New York City disappear?* 2024. URL: https://www.theguardian.com/global-development/2024/sep/26/how-did-an-argentina-lake-the-size-of-new-york-city-disappear-lake-colhue-huapi-musters-patagonia.

[331] NDIS Drought.gov. *Water Year 2024 Snow Drought Current Conditions Summary and Impacts in the West.* 2024. URL: https://www.drought.gov/drought-status-updates/water-year-2024-snow-drought-current-conditions-summary-and-impacts-west.

[332] Space News. *Lake Mead Keeps Dropping.* 2022. URL: https://spacenews.com/lake-mead-keeps-dropping/.

[333] *Mexico City Population 2025.* 2025. URL: https://worldpopulationreview.com/cities/mexico/mexico-city.

[334] Rodrigo Cervantes and Jérône Sessimi. *The Last Drops of Mexico City.* 2025. URL: https://mexicocitywater.longlead.com/.

[335] Carly Cassella. *Mexico City Is Sinking at an Alarming And Largely Unstoppable Rate, New Data Finds.* 2021. URL: https://www.sciencealert.com/mexico-city-is-sinking-at-an-alarming-and-largely-unstoppable-rate-according-to-data.

[336] Dimosthenis Vasiloudis. *'Chinampas': The Ancient Aztec Floating Gardens that hold promise for Future Urban Agriculture.* 2021. URL: https://www.thearchaeologist.org/blog/chinampas-the-ancient-aztec-floating-gardens-that-hold-promise-for-future-urban-agriculture.

[337] FAO — Land and Water Division. *Globally Important Agricultural Heritage Systems —- a legacy for the future.* URL: https://openknowledge.fao.org/server/api/core/bitstreams/0e0b56b6-1ac3-4258-a5b1-427949faed17/content.

[338] F. Yao et al. *Satellites reveal widespread decline in global lake water storage.* 2023. URL: https://www.science.org/doi/10.1126/science.abo2812.

[339] N. Hagelberg. *How climate change is making record-breaking floods the new normal.* 2020. URL: https://www.unep.org/news-and-stories/story/how-climate-change-making-record-breaking-floods-new-normal.

[340] N. Gunn and C. Rye. *A view of catastrophic flooding from across the world.* 2023. URL: https://www.wtwco.com/en-us/insights/2023/08/a-view-of-catastrophic-flooding-from-across-the-world.

[341] B. Henson and J. Masters. *Epic floods are wreaking havoc from Africa to Asia to Europe.* 2024. URL: https://yaleclimateconnections.org/2024/09/epic-floods-are-wreaking-havoc-from-africa-to-asia-to-europe/.

[342] The Weather Channel. *Report Reveals The Staggering Toll From Helene.* 2025. URL: https://weather.com/storms/hurricane/video/national-hurricane-center-final-report-helene.

[343] Global Environment Facility. *The importance of water sustainability.* 2015. URL: https://www.thegef.org/newsroom/news/importance-water-sustainability.

[344] Bernhardt L.M. *The interlinked threats facing lakes and why we need to protect them.* 2022. URL: https://www.unep.org/news-and-stories/story/interlinked-threats-facing-lakes-and-why-we-need-protect-them.

[345] Statista. *Water withdrawal shares worldwide by sector.* 2023. URL: https://www.statista.com/statistics/1258433/water-withdrawals-shares-worldwide-by-sector/.

[346] C. He et al. *Future global urban water scarcity and potential solutions.* 2021. URL: https://www.nature.com/articles/s41467-021-25026-3.

[347] dpopulationreview.com. *Desalination by Country 2024.* URL: https://worldpopulationreview.com/country-rankings/desalination-by-countrya.

[348] UNEP. *Five things to know about desalination.* 2021. URL: https://www.unep.org/news-and-stories/story/five-things-know-about-desalination.

[349] J. Chu. *Desalination system could produce freshwater that is cheaper than tap water.* 2023. URL: https://news.mit.edu/2023/desalination-system-could-produce-freshwater-cheaper-0927.

[350] J. Robbins. *In Seawater, Researchers See an Untapped Bounty of Critical Metals.* 2024. URL: https://e360.yale.edu/features/desalination-saltwater-brine-mining.

[351] press.un.org. *Bring the Water Action Agenda to Life.* 2023. URL: https://press.un.org/en/2023/sgsm21737.doc.htm.

[352] *Water Supply.* 2022. URL: https://www.worldbank.org/en/topic/watersupply.

[353] Unesco. *Statistics.* 2024. URL: https://www.unesco.org/reports/wwdr/en/2024/s.

[354] Global Commission on the Economics of Water. *The Economics of Water.* 2024. URL: https://watercommission.org/.

[355] One Tree Planted. *The 9 Oldest, Tallest, and Biggest Trees in the World.* 2024. URL: https://onetreeplanted.org/blogs/stories/oldest-tallest-biggest-trees?srsltid=AfmBOoq7M68W99RtZn3hfP-YPw8f6on4v3JhyzbCeU1U4v7mAsf2vzsV.

[356] National Park Service. *About the Trees - Redwood National and State Parks.* 2016. URL: https://www.nps.gov/redw/learn/nature/about-the-trees.htm.

[357] Sequoia & Kings Canyon National Parks. *General Sherman Tree — Sequoia National Park CA.* 2025. URL: https://www.visitsequoia.com/sequoia-national-park-attractions/general-sherman-tree/.

[358] Sequoia & Kings Canyon National Parks. *General Sherman Tree.* 2023. URL: https://www.nps.gov/places/000/general-sherman-tree.htm#:~:text=Rapid%5C%20growth%5C%20over%5C%20a%5C%20long,trees%5C%20maintain%5C%20balance%5C%20and%5C%20stability.

[359] The Whole World Is A Playground. *3 Incredible Drive Through Redwood Trees in California!* 2021. URL: https://www.thewholeworldisaplayground.com/drive-through-redwood-trees-california/.

[360] Chin A. *We Just Found a Secret Trait That May Help Redwood Trees Survive Climate Change.* 2022. URL: https://www.sciencealert.com/redwood-trees-have-a-secret-trait-that-helps-them-fight-climate-change-two-different-kinds-of-leaves.

[361] Drew MP Peltier and Andrew D Richardson. *fire, coast redwoods use decades-old carbon reserves to recover.* 2023. URL: https://www.nature.com/articles/s41477-023-01585-9.

[362] S. Quaglia. *Some redwood leaves make food while others drink water.* 2022. URL: https://www.snexplores.org/article/some-redwood-leaves-make-food-while-others-drink-water.

[363] E. Stokstad. *Ancient redwoods recover from fire by sprouting 1000-year-old buds.* 2023. URL: https://www.science.org/content/article/ancient-redwoods-recover-fire-sprouting-1000-year-old-buds.

[364] Guiness World Records. *Largest living tree (volume).* URL: https://www.guinnessworldrecords.com/world-records/largest-living-tree-.

[365] *How much do a Redwood Trees weigh and how much lumber do they contain?* 2021. URL: https://www.buyredwood.com/faq/how-much-do-redwood-trees-weigh-and-how-much-lumber-do-they-contain.

[366] Save the Redwoods League. *Coast Redwoods.* 2025. URL: https://www.savetheredwoods.org/redwoods/coast-redwoods/.

[367] Sierra Club. *History*. 2025. URL: https://www.sierraclub.org/maryland/chapter-history.

[368] Save the Redwoods League. *Mission and History*. 2025. URL: https://www.savetheredwoods.org/about-us/mission-history/.

[369] D. Mikkelson. *Ronald Reagan 'If You've Seen One Tree...'* 2006. URL: https://www.snopes.com/fact-check/if-youve-seen-one-tree/.

[370] Clarke Historical Museum. *Fight for the Redwoods*. 2025. URL: https://www.clarkemuseum.org/fight-for-the-redwoods.html.

[371] Save the Redwoods League. *Annual Report 2023-24: Reimagining redwood forests for the benefit of all*. 2025. URL: https://www.savetheredwoods.org/about-us/publications/2023-24-annual-report/.

[372] Sierra Club. *Forest Carbon Reserve*. 2025. URL: https://www.sierraclub.org/forests/forest-carbon-reserve.

[373] A. Harrell. *A California state forest is ground zero in a war over logging redwoods*. 2021. URL: https://www.sfgate.com/california-news/article/norcal-jackson-forest-redwood-logging-controversy-16530191.php.

[374] *State of the World's Mangroves*. 2021. URL: https://www.nature.org/en-us/what-we-do/our-insights/perspectives/state-of-world-mangroves/.

[375] A. Azoulay. *International Day for the Conservation of the Mangrove Ecosystem*. 2022. URL: https://www.unesco.org/en/days/mangrove-ecosystem-conservation.

[376] International Union for Conservation of Nature. *Red List of Mangrove Ecosystems*. 2024. URL: https://iucn.org/resources/conservation-tool/iucn-red-list-ecosystems/red-list-mangrove-ecosystems.

[377] Conservation International. *Mangroves*. URL: https://www.conservation.org/priorities/mangroves.

[378] J.W. Silk et al. *An estimate of the number of tropical tree species*. 2015. URL: https://www.pnas.org/doi/full/10.1073/pnas.1423147112.

[379] A. Nordseth. *Types of Forests: Definitions, Examples, and Importance*. 2022. URL: https://www.treehugger.com/types-of-forests-definitions-examples-5180645.

[380] WWF. *The Importance of Forests*. 2015. URL: https://wwf.panda.org/discover/our_focus/forests_practice/importance_forests/.

[381] H. Ritchie. *Forest area. Our World in Data*. 2021. URL: https://ourworldindata.org/forest-area.

[382] H. Richie. *The world has lost one-third of its forests, but an end to deforestation is possible.* 2025. URL: https://ourworldindata.org/world-lost-one-third-forests.

[383] J. MacCarthy et al. *New Data Confirms: Forest Fires Are Getting Worse.* 2023. URL: https://www.wri.org/insights/global-trends-forest-fires.

[384] ScienceDaily. *Climate change raised the odds of unprecedented wildfires in 2023-24.* 2023. URL: https://www.sciencedaily.com/releases/2024/08/240813192347.htm.

[385] Z. Budryk. *Emissions from Canada's 2023 wildfires higher than all but 3 countries: Study.* 2024. URL: https://www.yahoo.com/news/emissions-canada-2023-wildfires-higher-193523343.html.

[386] Harmsworth E. *Wildfires Cost Europe €4.1 Billion as Temperatures Hit Recordst.* 2023. URL: https://financialpost.com/pmn/business-pmn/greece-suffers-most-damage-as-wildfires-cost-europe-e4-1-billion.

[387] M. Heinrich. *Climate-exacerbated wildfires cost the U. S. between $394 to $893 billion each year in economic costs and damages.* 2023. URL: https://www.jec.senate.gov/public/index.cfm/democrats/2023/10/climate-exacerbated-wildfires-cost-the-u-s-between-394-to-893-billion-each-year-in-economic-costs-and-damages.

[388] Berardelli J. *Severe wildfires burning 8 times more area in western U.* 2020. URL: https://www.cbsnews.com/news/wildfires-western-united-states-8-times-more-area/.

[389] N. C. Dove, N. Taş, and S. C. Hart. *Ecological and genomic responses of soil microbiomes to high-severity wildfire: linking community assembly to functional potential.* 2022. URL: https://academic.oup.com/ismej/article/16/7/1853/7474254.

[390] I-Ching Chen et al. *Rapid range shifts of species associated with high levels of climate warming.* 2011. URL: https://www.science.org/doi/10.1126/science.1206432.

[391] P. Weston. *As the world gets hotter, could helping trees migrate northwards protect them? the Guardian.* 2024. URL: https://www.theguardian.com/environment/article/2024/aug/19/the-world-is-getting-hotter-could-helping-englands-trees-migrate-northwards-protect-them-climate-aoe.

[392] The UnitedNations. *Global Forest Goals and Targets of the UN Strategic Plan forForests 2030.* 2019. URL: https://www.un.org/esa/forests/wp-content/uploads/2019/04/Global-Forest-Goals-booklet-Apr-2019.pdf.

[393] UN News. *World leaders, corporations at COP26, take major step to restore and protect forests*. 2021. URL: https://news.un.org/en/story/2021/11/1104642.

[394] UN Environment Programme. *UN recognizes 10 pioneering initiatives that are restoring the natural world*. 2022. URL: https://www.unep.org/news-and-stories/press-release/un-recognizes-10-pioneering-initiatives-are-restoring-natural-world.

[395] Elizabeth Goldman and et al. Reytar Katie. *Deforestation and Restoration Targets Tracker (Beta)*. 2025. URL: https://gfr.wri.org/forest-targets-tracker#:~:text=Off.

[396] World Bank Groups. *Costa Rica's Forest Conservation Pays Off*. 2022. URL: https://www.worldbank.org/en/news/feature/2022/11/16/costa-rica-s-forest-conservation-pays-off.

[397] Matei Dolinay and Zuzana Dolinay. *Dave % Dave's Costra Rica Nature Park — hummingbird paradise*. 2018. URL: https://www.youtube.com/watch?v=j-0x92NiPVQ.

[398] The Economist. *Climate Change: can money stop deforestation?* 2023. URL: https://www.youtube.com/watch?v=cOfWyCkcSvg.

[399] World Bank Groups. *Costa Rica's Forest Conservation Pays Off*. 2022. URL: https://www.worldbank.org/en/news/feature/2022/11/16/costa-rica-s-forest-conservation-pays-off.

[400] *The State of World Fisheries and Aquaculture 2022*. 2023. URL: https://openknowledge.fao.org/server/api/core/bitstreams/9df19f53-b931-4d04-acd3-58a71c6b1a5b/content/sofia/2022/consumption-of-aquatic-foods.html.

[401] A. Vaughan. *Global demand for fish expected to almost double by 2050*. 2021. URL: https://www.newscientist.com/article/2290082-global-demand-for-fish-expected-to-almost-double-by-2050/.

[402] Food and Agriculture Orginazation of the United Nations. *The State of World Fisheries and Aquaculture*. 2024. URL: https://openknowledge.fao.org/server/api/core/bitstreams/51cd4687-66bf-402e-94b8-eb963d6a8b00/content.

[403] Bennett A., Patil P, and et al. Kleisner K. *Contribution of Fisheries to Food and Nutrition Security: Current Knowledge, Policy, and Research*. 2018. URL: https://wfpc.sanford.duke.edu/reports/contribution-fisheries-food-nutrition-security-current-knowledge-policy-and-research/.

[404] C. Palmer. *Marine Biodiversity and Ecosystems Underpin a Healthy Planet and Social Well-Being*. 2017. URL: https://www.un.org/en/chronicle/article/marine-biodiversity-and-ecosystems-underpin-healthy-planet-and-social-well-being.

[405] C. D. Golden et al. *Nutrition: Fall in fish catch threatens human health*. 2016. URL: https://www.nature.com/articles/534317a.

[406] H. Richie. *One-third of the world's assessed fish stocks are overexploited*. 2024. URL: https://ourworldindata.org/data-insights/one-third-of-the-worlds-assessed-fish-stocks-are-overexploited#:~:text=One%5C%2Dthird%5C%20of%5C%20the%5C%20world's,overexploited%5C%20%5C%2D%5C%20Our%5C%20World%5C%20in%5C%20Data.

[407] *The Modern Day Pacific Sardine Collapse: How to Prevent a Future Crisis*. 2018. URL: https://usa.oceana.org/responsible-fishing-modern-day-pacific-sardine-collapse-how-prevent-future-crisis/.

[408] OCEANA. *New Catch Limits Issured for Northern Anchovy Still Violate the Law*. 2024. URL: %5Chttps://usa.oceana.org/press-releases/new-catch-limits-issued-northern-anchovy-still-violate-law/.

[409] M. Moyer and C. Storrs. *How Much Is Left?* 2010. URL: https://www.scientificamerican.com/article/how-much-is-left/.

[410] L. Kubiak. *Marine Biodiversity in Dangerous Decline*. 2019. URL: https://www.nrdc.org/bio/lauren-kubiak/marine-biodiversity-dangerous-decline-finds-new-report.%20Finds%20New%20Report.%20www.nrdc.org.

[411] G. Gustin. *Climate Change Threatens the World's Fisheries, Food Billions of People Rely On*. 2019. URL: https://insideclimatenews.org/news/29092019/ocean-fish-diet-climate-change-impact-food-ipcc-report-cryosphere/.%20G.%20Climate%20Change%20Threatens%20the%20World's%20Fisheries,%20Food%20Billions%20of%20People%20Rely%20Ons.

[412] G. Readfearn. *Hottest ocean temperatures in 400 years an 'existential threat' to the Great Barrier Reef, researchers find. the Guardian*. 2024. URL: https://www.theguardian.com/environment/article/2024/aug/08/great-barrier-reef-ocean-temperatures-hottest-record.

[413] D. Anguiano. *Florida ocean records 'unprecedented' temperatures similar to a hot tub*. 2023. URL: https://www.theguardian.com/us-news/2023/jul/25/florida-ocean-temperatures-hot-tub-extreme-weather.

[414] T. Dong, Z. Zeng, and et al. Wang D. *Record-breaking 2023 marine heatwaves*. 2025. URL: https://www.science.org/doi/10.1126/science.adr0910?utm%5C_source=sfmc%5C&utm%5C_medium=email%5C&utm%5C_campaign=ScienceAdviser%5C&utm%5C_content=distillation%5C&et%5C_rid=17081200%5C&et%5C_cid=5686989.

[415] Copernicus Climate Change Service. *Global Climate Highlights 2024*. 2024. URL: https://climate.copernicus.eu/global-climate-highlights-2024.

[416] Microplastics. *are everywhere — we need to understand how they affect human health*. 2024. URL: https://www.nature.com/articles/s41591-024-02968-x.

[417] A. Alfaro-Núñez et al. *Microplastic pollution in seawater and marine organisms across the Tropical Eastern Pacific and Galápagos*. 2021. URL: https://www.nature.com/articles/s41598-021-85939-3.

[418] M. K. Dziobak et al. *First evidence of microplastic inhalation among free-ranging small cetaceans*. 2024. URL: https://journals.plos.org/plosone/article?id=10.1371/journal.pone.0309377.

[419] S. Dutchen. *Microplastics Everywhere*. 2025. URL: https://magazine.hms.harvard.edu/articles/microplastics-everywhere.

[420] J. Kang, L. Zhou, and et al. Sun H. *Degradation of Cosmetic Microplastics via Functionalized Carbon Nanosprings*. 2019. URL: https://www.cell.com/matter/fulltext/S2590-2385(19)30056-6.

[421] C. Lim et al. *Bioplastic made from seaweed polysaccharides with green production methods*. 2021. URL: https://www.sciencedirect.com/science/article/abs/pii/S2213343721008721.

[422] NSF Stories. *Genetically modified bacteria break down plastics in saltwater*. 2023. URL: https://www.nsf.gov/news/genetically-modified-bacteria-break-down-plastics#:~:text=Modified%20organism%20can%20break%20down,saltwater%20environment%20at%20room%20temperature.

[423] UNESCO. *United Nations Decade of Ocean Science for Sustainable Development (2021–2030)*. 2021. URL: United%20Nations%20Decade%20of%20Ocean%20Science%20for%20Sustainable%20Development.

[424] F. Robinet. *UN ocean summit in Nice closes with wave of commitments*. 2025. URL: https://news.un.org/en/story/2025/06/1164381.

[425] Convention on Biological Diversity. *Kumming-Montreal Global Biodivesity Framework*. 2022. URL: https://www.unep.org/resources/kunming-montreal-global-biodiversity-framework.

[426] Bloomberg Philanthropies Ocean Initiative. *On track or off course?* 2024. URL: https://assets.bbhub.io/dotorg/sites/64/2024/10/On-track-or-off-course.pdf.

[427] United Nations Environment Programme. *What is the international agreement to protect the high seas and why is it important?* 2025. URL: https://www.unep.org/news-and-stories/story/what-international-agreement-protect-high-seas-and-why-it-important.

[428] High Seas Alliance. *HISTORIC MILESTONE FOR GLOBAL OCEAN PROTECTION.* 2025. URL: https://highseasalliance.org/2025/09/19/historic-milestone-for-global-ocean-protection-60th-ratification-triggers-entry-into-force-of-high-seas-treaty/.

[429] *High Level Panel for a Sustainable Ocean Economy.* 2025. URL: https://oceanpanel.org/.

[430] World Resources Institute. *Ocean Action 2030.* 2025. URL: https://www.wri.org/initiatives/ocean-action-2030.

[431] M. Mussi, M. M. Skinner, and et al. Melis R. T. *Early Homo erectus lived at high altitudes and produced both Oldowan and Acheulean tools.* 2023. URL: https://www.science.org/doi/10.1126/science.add9115.

[432] E. Wayman. *The Earliest Example of Hominid Fire.* 2012. URL: https://www.smithsonianmag.com/science-nature/the-earliest-example-of-hominid-fire-171693652/.

[433] B. Handwerk. *Two Million Years Ago, This Homo Erectus Lived the High Life.* 2023. URL: https://www.smithsonianmag.com/science-nature/two-million-years-ago-homo-erectus-likely-lived-in-cool-high-altitude-environments-180983073/.

[434] B. Handwerk. *Fossils From Some of the Last Homo Erectus Hint at the End of the Long-Lived Species.* 2019. URL: https://www.smithsonianmag.com/science-nature/fossils-some-last-homo-erectus-hint-end-long-lived-species-180973816/.

[435] *The First Hunter-Gatherers – The Story of the Human Body: Evolution, Health, and Disease.* 2020. URL: https://sites.nd.edu/caroline-langley/2020/04/25/the-first-hunter-gatherers/.

[436] S. Naithani. *The Origins of Agriculture. open.oregonstate.education.* 2021. URL: https://open.oregonstate.education/cultivatedplants/chapter/agriculture/.

[437] Richard B Lee. *The !Kung San : men, women, and work in a foraging society.* 1979. URL: https://archive.org/details/kungsanmenwomenwo000leer.

[438] Padilla-Iglesias. C. *The hunter-gatherers of the 21st century who live on the move*. 2024. URL: https://aeon.co/essays/the-hunter-gatherers-of-the-21st-century-who-live-on-the-move.

[439] J. Fassler. *What Some of the World's Last Hunter-Gathers Have to Say*. 2019. URL: https://www.theatlantic.com/entertainment/archive/2019/02/what-doug-bock-clark-learned-last-whalers/582643/.

[440] *First evidence of farming in Mideast 23,000 years ago: Evidence of earliest small-scale agricultural cultivation*. 2015. URL: https://www.sciencedaily.com/releases/2015/07/150722144709.htm.

[441] A. A. Macintosh, R. Pinhasi, and J. T. Stock. *Prehistoric women's manual labor exceeded that of athletes through the first 5500 years of farming in Central Europe*. https://www.science.org/doi/10.1126/sciadv.aao3893. 2017.

[442] F. Diep. *8,000 Years Ago, 17 Women Reproduced for Every One Man*. 2015. URL: https://psmag.com/environment/17-to-1-reproductive-success/.

[443] M. Karmin et al. *A recent bottleneck of Y chromosome diversity coincides with a global change in culture*. 2015. URL: https://genome.cshlp.org/content/25/4/459.

[444] J. Hancock. *Origins of World Agriculture*. 2021. URL: https://www.worldhistory.org/article/1886/origins-of-world-agriculture/.

[445] M. Melletti. *Cattle Domestication: from Aurochs to Cow - Fifteen Eighty Four*. 2016. URL: https://cambridgeblog.org/2016/02/cattle-domestication-from-aurochs-to-cow/.

[446] Hirst K. *How the Wild Boar Became the Sweet Domestic Pig*. 2019. URL: https://www.thoughtco.com/the-domestication-of-pigs-170665.

[447] *New light shed on the domestication history of sheep and goats*. 2018. URL: https://www.cardiff.ac.uk/news/view/1128739-new-light-shed-on-the-domestication-history-of-sheep-and-goats.

[448] Hirst K. *Who Gets the Credit for Taming the Wild Jungle Fowl? ThoughtCo*. 2019. URL: https://www.thoughtco.com/the-domestication-history-of-chickens-170653.

[449] The Rice Association. *History of rice*. URL: https://www.riceassociation.org.uk/history-of-rice.

[450] V. H. Adler. *Hunter Gatherers*. URL: https://bio.libretexts.org/Courses/Coalinga_College/Introduction_to_Plant_Science_(Hochman_Adler)/02%5C%3A_Importance_of_plants/2.01%5C%3A_The_Origins_of_Agriculture.

[451] *History of Chocolate*. 2017. URL: https://www.history.com/topics/ancient-americas/history-of-chocolate.

[452] E. Malaker. *The Fascinating History of Coffee: Where Did Coffee Originate?* 2022. URL: https://coffeeaffection.com/history-of-coffee/.

[453] Max_zero. *The Full History of Agriculture: From Hunter-Gatherers to Modern Farming*. 2023. URL: https://agtecher.com/the-full-history-of-agriculture-from-hunter-gatherers-to-modern-farming/.

[454] *Soil Basics, Soil Science Society of America*. 2024. URL: https://www.soils.org/about-soils/basics/.

[455] *FAO warns 90 per cent of Earth's topsoil at risk by 2050*. 2022. URL: https://news.un.org/en/story/2022/07/1123462.

[456] *Soil pollution 'jeopardizing' life on Earth, UN agency warns on World Day*. 2018. URL: https://news.un.org/en/story/2018/12/1027681.

[457] John Todd. *Healing Earth: An Ecologist's Journey of Innovation and Environmental Stewardship*. 2019. URL: https://www.penguinrandomhouse.com/books/576932/healing-earth-by-john-todd/.

[458] M. Schmidt et al. *Intentional creation of carbon-rich dark earth soils in the Amazon*. 2023. URL: https://www.science.org/doi/10.1126/sciadv.adh8499.

[459] *This man-made super-soil is how ancient Amazon civilizations bio-designed 12% of the rainforest*. 2015. URL: https://theplaidzebra.com/this-man-made-super-soil-is-how-ancient-amazon-civilizations-bio-designed-12-of-the-rainforest/.

[460] Apte A. *Digging Deep: Investigating Manmade Black Soil of the Amazon, Abdul Latif Jameel Water and Food Systems Lab (J-WAFS)*. 2019. URL: https://jwafs.mit.edu/news/2019/digging-deep-investigating-manmade-black-soil-amazon.

[461] H. Ritchie and M. Roser. *Half of the world's habitable land is used for agriculture. Our World in Data*. 2024. URL: https://ourworldindata.org/global-land-for-agriculture.

[462] *The state of the world's land and water resources for food and agriculture, Land & Water*. 2024. URL: https://www.fao.org/land-water/solaw2021/en.

[463] *STATISTICAL YEARBOOK WORLD FOOD AND AGRICULTURE 2023*. 2023. URL: https://openknowledge.fao.org/server/api/core/bitstreams/28cfd24e-81a9-4ebc-b2b5-4095fe5b1dab/content/cc8166en.html/chapter-1.

[464] D. Nuccitelli. *UN report: The world's farms stretched to 'a breaking point"' Yale Climate Connections*. 2022. URL: https://yaleclimateconnections.org/2022/01/un-report-the-worlds-farms-stretched-to-a-breaking-point/.

[465] O'Brian D. *Food Banks Provided Vital Services to Meet Persistently High Demand in 2023, The Global FoodBanking Network*. 2024. URL: https://www.foodbanking.org/blogs/food-banks-provided-vital-services-to-meet-high-demand/.

[466] H. Ritchie. *How many animals are factory-farmed?* 2023. URL: https://ourworldindata.org/how-many-animals-are-factory-farmed.

[467] A. Quarracino. *12 Fact About Factory Farming and the Environment*. 2022. URL: https://thehumaneleague.org/article/12-facts-about-factory-farming-and-the-environment.

[468] J. Anomaly. *What's Wrong With Factory Farming?* 2014. URL: https://pmc.ncbi.nlm.nih.gov/articles/PMC9757169/.

[469] *Intensive poultry farming: A review of the impact on the environment and human health*. 2023. URL: https://www.sciencedirect.com/science/article/pii/S0048969722071145.

[470] A. Schechinger. *New USDA data reveal that the largest factory farms keep growing in number, Environmental Working Group*. 2024. URL: https://www.ewg.org/news-insights/news/2024/02/new-usda-data-reveal-largest-factory-farms-keep-growing-number.

[471] C. Buckley. *Life After Factory Farming: 'The Longer They're Out, the Happier They Are.' Nytimes.com*. 2024. URL: https://www.nytimes.com/2024/08/29/climate/factory-farming-transfarmation-project.html.

[472] H. Decker. *What Is Contract Farming?* 2024. URL: https://thetransfarmationproject.org/blog/what-is-contract-farming/.

[473] C. Buckley. 2024. URL: Ubid.

[474] Farmdoc. *Farm Bankruptcies This Year Already Exceed 2024 Levels*. 2025. URL: https://www.agriculture.com/partners-farm-bankruptcies-this-year-already-exceed-2024-levels-11772290.

[475] *What Is Sustainable agriculture?* 2022. URL: https://www.ucsusa.org/resources/what-sustainable-agriculture.

[476] J. Ranganathan et al. *How to Sustainably Feed 10 Billion People by 2050 in 21 Charts*. 2018. URL: https://www.wri.org/insights/how-sustainably-feed-10-billion-people-2050-21-charts.

[477] *How Gender Justice Can Advance Climate Resilience and Zero Hunger*. 2023. URL: https://www.globalhungerindex.org/.

[478] Union of Concerned Scientists. *What Is Sustainable Agriculture?* 2022. URL: https://www.ucs.org/resources/what-sustainable-agriculture.

[479] C. Buckley. *Much of Ireland Is an Ecological Desert. Meet the Man Who Wants to Rewild It*. 2024. URL: https://www.nytimes.com/2024/10/18/climate/ireland-ecological-desert-rewilding.html.

[480] S. Lasley. *USGS expands critical minerals list to 54*. 2025. URL: https://www.metaltechnews.com/story/2025/08/27/tech-metals/usgs-expands-critical-minerals-list-to-54/2451.html.

[481] Merrill A. *Aluminum Statistics and Information, U. S.* 2024. URL: https://www.usgs.gov/centers/national-minerals-information-center/aluminum-statistics-and-information.

[482] *Zirconium: resources, reserves and production-Metalpedia*. 2014. URL: http://metalpedia.asianmetal.com/metal/zirconium/resources%5C&production.shtml.

[483] IEA. *Mineral requirements for clean energy transitions – The Role of Critical Minerals in Clean Energy Transitions*. 2021. URL: https://www.iea.org/reports/the-role-of-critical-minerals-in-clean-energy-transitions/mineral-requirements-for-clean-energy-transitions.

[484] IEA 50. *Clean energy demand for critical minerals set to soar as the world pursues net zero goals*. 2021. URL: https://www.iea.org/news/clean-energy-demand-for-critical-minerals-set-to-soar-as-the-world-pursues-net-zero-goals.

[485] scripts.mit.edu. *Mining, Water for all*. 2017. URL: %5Craggedrighthttp://12.000.scripts.mit.edu/mission2017/mining/.

[486] Provident Metals. *Geology of Platinum*. 2024. URL: https://www.providentmetals.com/knowledge-center/precious-metals-resources/geology-of-platinum.html.

[487] C. Kousky et al. *China's role in supplying critical minerals for the global energy transition: What could the future hold? Brookings*. 2022. URL: https://www.brookings.edu/articles/chinas-role-in-supplying-critical-minerals-for-the-global-energy-transition-what-could-the-future-hold/.

[488] S. S. Zhang et al. *China Slaps Export Restrictions on Two Critical Metals*. 2023. URL: https://www.globalpolicywatch.com/2023/07/china-slaps-export-restrictions-on-two-critical-metals/.

[489] FinancialTimes. *Western nations join forces to break China's grip on critical minerals*. 2018. URL: https://www.ft.com/content/2984ae03-df15-420b-89cc-9ad8337014a9.

[490] *Ending Exploitation in the DRC's Deadly Cobalt Mines*. 2023. URL: https://globalhumanrights.org/stories/extraction-without-exploitation/.

[491] E. Silva. *Indonesian nickel production dominates commodity market. www.spglobal.com*. 2024. URL: https://www.spglobal.com/marketintelligence/en/news-insights/latest-news-headlines/indonesian-nickel-production-dominates-commodity-market-80242322.

[492] J. Rachman. *US may block Indonesia nickel on forced labor issues*. 2024. URL: https://asiatimes.com/2024/09/us-may-block-indonesia-nickel-on-forced-labor-issues/.

[493] L. David. *Moon mining gains momentum as private companies plan for a lunar economy*. 2023. URL: https://www.space.com/moon-mining-gains-momentum.

[494] M. Hutson. *The Quest to Build a Telescope on the Moon*. 2024. URL: https://www.newyorker.com/science/elements/the-quest-to-build-a-telescope-on-the-moon.

[495] B. A. Sharkh et al. *Seawater desalination concentrate—a new frontier for sustainable mining of valuable minerals*. 2022. URL: https://www.nature.com/articles/s41545-022-00153-6.

[496] T. De Chant. *Tidal Metals sees seawater as the solution to a critical mineral shortage*. 2024. URL: https://www.yahoo.com/news/tidal-metals-sees-seawater-solution-140000616.html

[497] I. Morse. *Down on the Farm That Harvests Metal From Plants*. 2020. URL: https://www.nytimes.com/2020/02/26/science/metal-plants-farm.html.

[498] R. McKie. *Bacteria helping to extract rare metals from old batteries in boost for green tech. the Guardian*. 2024. URL: https://www.theguardian.com/environment/article/2024/aug/25/bacteria-helping-to-extract-rare-metals-from-old-batteries-in-boost-for-green-tech.

[499] *Deep-Sea Mining Could Help Meet Demand for Critical Minerals, But Also Comes with Serious Obstacles*. 2021. URL: https://www.gao.gov/blog/deep-sea-mining-could-help-meet-demand-critical-minerals.

[500] D. McCauley and D. Amon. *Risks of deep-sea mining are not fully understood: Here's why that matters.* 2022. URL: https://www.weforum.org/agenda/2022/07/take-a-deep-dive-into-how-deep-sea-mining-could-threaten-our-oceans/.

[501] J. Jackson and L. Karn. *Seabed Mining Moratorium Is Legally Required by U.* 2023. URL: https://www.pewtrusts.org/en/research-and-analysis/fact-sheets/2023/06/seabed-mining-moratorium-is-legally-required-by-un-treaty-legal-experts-find.

[502] Z. Hu et al. *Circular economy strategies for mitigating metals shortages in electric vehicle batteries under China's carbon-neutral target.* https://www.sciencedirect.com/science/article/abs/pii/S0301479724000653?via%3Dihub. 2024.

[503] P. Ferro, F. Bonollo, and S. Cruz. *Alloy Substitution in a Critical Raw Materials Perspective.* 2019. URL: https://www.fracturae.com/index.php/fis/article/view/2595.

[504] A. Guterres. *The UN Secretary-General's Panel on Critical Energy Transition Minerals.* 2024. URL: https://www.un.org/en/climatechange/critical-minerals.

[505] Ellen MacArthur Foundation. *Schools of thought that inspired the circular economy. www.ellenmacarthurfoundation.org.* 2023. URL: https://www.ellenmacarthurfoundation.org/schools-of-thought-that-inspired-the-circular-economy.

[506] Ellen Macarthur Foundation. *The Circular Economy in Detail.* 2024. URL: https://www.ellenmacarthurfoundation.org/the-circular-economy-in-detail-deep-dive.

[507] D. McGinty. *5 Opportunities of a Circular Economy.* 2021. URL: https://www.wri.org/insights/5-opportunities-circular-economy.

[508] Circle Economy Foundation. *Circular economy strategies can cut global emissions by 39%.* 2024. URL: https://www.circle-economy.com/news/circular-economy-strategies-can-cut-global-emissions-by-39.

[509] www.circularity-gap.world. *The Circularity Gap Report 2024.* 2024. URL: https://www.circularity-gap.world/2024.

[510] J. Vaillant. *We built our world with fire. Now heat is destroying our lives.* 2025. URL: https://www.theguardian.com/us-news/2025/jan/11/we-built-our-world-with-fire-now-heat-destroying-our-lives.

[511] Monica Danielle. *AccuWeather estimates more than $250 billion in damages and economic loss from LA wildfires*. 2025. URL: https://www.accuweather.com/en/weather-news/accuweather-estimates-more-than-250-billion-in-damages-and-economic-loss-from-la-wildfires/1733821.

[512] Anon. *Extreme weather caused two million deaths, cost $4 trillion over last 50 years*. 2023. URL: https://news.un.org/en/story/2023/05/1136897.

[513] Richard Heinberg. *Museletter #380: Envisioning a Livable Future*. 2024. URL: https://richardheinberg.com/museletter-380-envisioning-a-livable-future.

[514] Mason Yarbrough. *Engineered Addictions*. 2025. URL: https://masonyarbrough.substack.com/p/engineered-addictions.

[515] M. Igini. *11 Inspiring Jane Goodall Quotes About Our Planet's Future*. 2025. URL: https://earth.org/jane-goodall-quotes/.

[516] Anon. *Executive summary – World Energy Employment 2023*. URL: https://www.iea.org/reports/world-energy-employment-2023/executive-summary.

[517] Anon. *Fossil Fuel Investment Report*. URL: https://investinginclimatechaos.org/reports.

[518] Richie H. *Solar panel prices have fallen by around 20% every time global capacity doubled*. URL: https://ourworldindata.org/data-insights/solar-panel-prices-have-fallen-by-around-20-every-time-global-capacity-doubled.

[519] Anon. *Trends in electric cars – Global EV Outlook 2024 – Analysis*. URL: https://www.iea.org/reports/global-ev-outlook-2024/trends-in-electric-cars.

[520] UN Trade and Development. *The costs of achieving the Sustainable Development Goals*. 2023. URL: https://unctad.org/sdg-costing.

[521] United Nations. *Five ways to jump-start the renewable energy transition now*. URL: https://www.un.org/en/climatechange/raising-ambition/renewable-energy-transition

[522] PennState Institute of Energy and the Environment. *Transitioning to renewable energy: Challenges and opportunities*. 2024. URL: https://iee.psu.edu/news/blog/transitioning-renewable-energy-challenges-and-opportunities.

[523] Cara Goldenberg and et al. Gomez-Contreras Anibal. *Fossil Fuel Transition Strategies*. 2022. URL: https://rmi.org/insight/global-energy-transformation-guide-fossil-fuel-transition-strategies/.

[524] Companies Marketcap. *Largest oil and gas companies by market cap.* 2025. URL: https://companiesmarketcap.com/oil-gas/largest-oil-and-gas-companies-by-market-cap/.

[525] P. Denholm, P. Brown, and et al. Cole W. *Examining Supply-Side Options to Achieve 100% Clean Energy by 2035.* 2022. URL: https://docs.nrel.gov/docs/fy22osti/81644.pdf.

[526] T. Vonyá. *Recovery and reconstruction: Europe after WWII.* 2019. URL: https://cepr.org/voxeu/columns/recovery-and-reconstruction-europe-after-wwii.

[527] National Museum of American Diplomacy. *The Marshall Plan - The National Museum of American Diplomacy.* 2025. URL: https://diplomacy.state.gov/online-exhibits/diplomacy-is-our-mission/development/the-marshall-plan/.

[528] InfluenceMap. *Big Oil's Real Agenda on Climate Change 2022.* URL: https://influencemap.org/report/Big-Oil-s-Agenda-on-Climate-Change-2022-19585.

[529] Jeffery Burns. *Sustainable Agriculture Trends: What's Shaping the Future of Farming?* 2025. URL: https://sustainable-usa.org/sustainable-agriculture-trends-what-s-shaping-the-future-of-farming/.

[530] Clare P. *Overview — worldbank.org.* URL: https://www.worldbank.org/en/topic/poverty/overview.

[531] et al. Lowe Ian. *The Population Debate Revisited.* 2022. URL: https://tellus.org/wp-content/uploads/2022/08/Population-Debate-Revisited.pdf.

[532] Natalia V Bhattacharjee and et al. Schumacher Austin E. *Global fertility in 204 countries and territories, 1950–2021, with forecasts to 2100: a comprehensive demographic analysis for the Global Burden of Disease Study 2021.* 2024. URL: https://www.thelancet.com/journals/lancet/article/PIIS0140-6736(24)00550-6/fulltext.

[533] World Population Review. *Total Fertility Rate 2025.* 2025. URL: https://worldpopulationreview.com/country-rankings/total-fertility-rate.

[534] Liam McBain. *Why are people freaking out about the birth rate?* 2025. URL: https://www.npr.org/2025/06/27/nx-s1-5416788/birth-rate-pronatalism-culture.

[535] G. Chandler. *10 facts about Greta Thunberg.* URL: https://www.natgeokids.com/nz/kids-club/cool-kids/general-kids-club/greta-thunberg-facts/.

[536] *Fridays for Future...* 2025. URL: https://fridaysforfuture.org/what-we-do/who-we-are/.

[537] UNESCO. *Youth Climate Action Network.* URL: https://www.unesco.org/en/youth/climate-action-network.

[538] B. Jarvis. *How a Group of Students in the Pacific Islands Reshaped Global Climate Law.* 2025. URL: https://www.nytimes.com/2025/09/10/magazine/global-climate-law-students.html.

[539] Center for International Environmental Law. *Legal Memorandum Advisory Opinion on Climate Change Delivered by the International Tribunal for the Law of the Sea: Relevance for the International Court of Justice Climate Advisory Proceedings (July 2024).* 2024. URL: https://www.ciel.org/reports/legal-memorandum-advisory-opinion-on-climate-change-itlos/.

[540] United Nations Human Rights. *UN experts hail landmark Inter-American Court opinion on States' extensive duties to protect the climate.* 2025. URL: https://www.ohchr.org/en/press-releases/2025/07/un-experts-hail-landmark-inter-american-court-opinion-states-extensive.

[541] Climate Change Litigation Databases. *Request for an advisory opinion on the human rights obligations of African states in addressing the climate crisis.* 2025. URL: https://climatecasechart.com/non-us-case/request-for-an-advisory-opinion-on-the-human-rights-obligations-of-african-states-in-addressing-the-climate-crisis/.

[542] Chinese Academy of Sciences. *Weather and climate extremes in 2023 impacting the globe with emerging features.* 2025. URL: https://phys.org/news/2024-04-weather-climate-extremes-impacting-globe.pdf.

[543] Hanna C. *Humanity has warmed the planet by 1.5 ° C since 1700.* 2020. URL: https://www.newscientist.com/article/2455715-humanity-has-warmed-the-planet-by-1-5c-since-1700/.

[544] Exxon Knew. *about Climate Change Almost 40 years ago.* 2016. URL: https://www.scientificamerican.com/article/exxon-knew-about-climate-change-almost-40-years-ago/.

[545] G. McBean. *The oil and gas industry has been lying about global warming for decades — accountability is long overdue.* 2024. URL: https://theconversation.com/the-oil-and-gas-industry-has-been-lying-about-global-warming-for-decades-accountability-is-long-overdue-230160.

[546] C. Owen-Burge. *The IPCC just published its summary of 5 years of reports – here's what you need to know*. 2023. URL: https://climatechampions.unfccc.int/the-ipcc-just-published-its-summary-of-5-years-of-reports-heres-what-you-need-to-know/.

[547] Unfccc.int. 2022. URL: https://unfccc.int/news/everything-you-need-to-know-about-the-ipcc-report.

[548] A. Hoekstra. *Batteries: how cheap can they get?* 2024. URL: https://aukehoekstra.substack.com/p/batteries-how-cheap-can-they-get.

[549] K. Raworth. *Doughnut economics: seven ways to think like a 21st-century economist*. London: Random House Business Books, 2017.

[550] Forbes. *Billionaires 2024*. 2024. URL: https://www.forbes.com/billionaires/.

[551] Informaconnect.com. *Companies Entering The Trillion Dollar Club in 2024, Informa Connect Academy*. 2024. URL: https://informaconnect.com/academy/companies-entering-trillion-dollar-club-in-2024/.

[552] USICH. *United States Interagency Council on Homelessness (USICH)*. 2019. URL: https://www.usich.gov/.%20Usich.gov.

[553] S. Levin and W. Craft. *Thousands of senior citizens are dying homeless in Los Angeles, records reveal. the Guardian*. 2024. URL: https://www.theguardian.com/us-news/article/2024/aug/08/homeless-senior-citizen-deaths-los-angeles.

[554] Incorporated Industrial Economics. *Impact of extreme heat to California's people, infrastructure, and economy*. 2024. URL: https://www.insurance.ca.gov/01-consumers/180-climate-change/upload/Impacts-of-extreme-heat-to-California-s-people-infrastructure-and-economy-by-California-Department-of-Insurance-June-2024.pdf.

[555] I. D. Yosef, Tim Lister, and Eugenia Khader Al-Za'anoun. *Almost entire population in Gaza now displaced amid fresh Israeli offensive*. 2024. URL: https://www.cnn.com/2024/07/06/middleeast/palestinians-displaced-gaza-israel-intl/index.html.

[556] Anon. *Birth Rate By Country 2025 — datapandas.org*. 2025. URL: https://www.datapandas.org/ranking/birth-rate-by-country.

[557] *ibid.*

[558] M. Igini. *The Tipping Points of Climage Change: How will Our World Change?* 2024. URL: https://earth.org/tipping-points-of-climate-change/.

[559] US EPA. *Resources, Waste and Climate Change*. 2021. URL: https://www.epa.gov/smm/resources-waste-and-climate-change.

[560] Trading Economics. *United States GDP per capita*. 2024. URL: https://tradingeconomics.com/united-states/gdp-per-capita-ppp.

[561] Trading Economics. *Burundi GDP per capita*. 2023. URL: https://tradingeconomics.com/burundi/gdp.

[562] A. Kapoor and B. Debroy. *GDP Is Not a Measure of Human Well-Being*. 2019. URL: https://hbr.org/2019/10/gdp-is-not-a-measure-of-human-well-being.

[563] M. A. Killingsworth. *Money and Happiness: Extended Evidence Against Satiation | Happiness Science*. 2024. URL: https://happiness-science.org/money-happiness-satiation/.

[564] Anon. *Dave Ramsey — Wikipedia*. 2025. URL: https://en.wikipedia.org/wiki/Dave_Ramsey.

[565] K. R. Miner et al. *Permafrost carbon emissions in a changing Arctic*. 2022. URL: https://www.nature.com/articles/s43017-021-00230-3.

[566] Westen Emma. *Probability Estimates of a 21st Century AMOC Collapse*. 2024. URL: https://arxiv.org/abs/2406.11738.

[567] Timothy Lenfon, David I. Armstrong McKay, and et al. Soriani Sara. *Global Tipping Points*. 2023. URL: file:///home/lloyd/Downloads/GTP-full-report-071223.pdf.

[568] Tara Laan, Anna Geddes, and et al. Do Nhat. *Burning Billions: Record Public Money for Fossil Fuels Impeding Climate Action*. 2022. URL: https://www.energypolicytracker.org/burning-billions-record-fossil-fuels-support-2022/.

[569] L. de Nadal. *From Denial to the Culture Wars: A Study of Climate Misinformation on YouTube*. 2024. URL: https://www.tandfonline.com/doi/full/10.1080/17524032.2024.2363861.

[570] meduza.io. *like I drew a door and disappeared through it*. 2024. URL: https://meduza.io/en/feature/2021/04/01/it-s-like-i-drew-a-door-and-disappeared-through-it.

[571] Yahoo Finance. *How private equity firms make money*. 2021. URL: https://finance.yahoo.com/news/private-equity-firms-money-050000580.html.

[572] A Bastish. *Equilar / New York Times 100 Highest-Paid CEOs*. 2024. URL: https://www.equilar.com/reports/111-equilar-new-york-times-top-100-highest-paid-ceos-2024.html.

[573] L. Akabas. *Highest Paid Athletes in the World: Full List*. 2024. URL: https://www.sportico.com/feature/highest-paid-athletes-in-the-world-1234765608/.

[574] Forbes. *The World's Highest-Paid Celebrities*. 2020. URL: https://www.forbes.com/celebrities/list/.

[575] Sebastian Blanco. *Report: Tesla Autopilot Involved in 736 Crashes since 2019*. 2023. URL: https://www.caranddriver.com/news/a44185487/report-tesla-autopilot-crashes-since-2019/.

[576] Kate Whiting. *This is how AI is impacting – and shaping – the creative industries, according to experts at Davos*. 2024. URL: https://www.weforum.org/stories/2024/02/ai-creative-industries-davos/.

[577] United States Geological Survey. *What is remote sensing and what is it used for?* 2025. URL: https://www.usgs.gov/faqs/what-remote-sensing-and-what-it-used.

[578] J. Murphy, M. Roser, and E. Ortiz-Ospina. *Internet*. 2023. URL: https://ourworldindata.org/internet.

[579] E. Charlton. *2023 was a big year for cybercrime – here's how we can make our systems safer*. 2024. URL: https://www.weforum.org/stories/2024/01/cybersecurity-cybercrime-system-safety/.

[580] *UN Secretary-General launches UN Global Principles for Information Integrity*. 2024. URL: https://caribbean.un.org/en/272440-un-secretary-general-launches-un-global-principles-information-integrity.

[581] *Natural Resources: Why Are Natural Resources at Risk of Illicit Financial Flows?* 2025. URL: https://gfintegrity.org/issue/natural-resources/.

[582] *Sustainability Development Goals: 17 Goals to Transform our World*. URL: https://www.un.org/en/exhibits/page/sdgs-17-goals-transform-world.

[583] *Kate Raworth*. 2019. URL: https://en.wikipedia.org/wiki/Kate%5C_Raworth.

[584] P. Leahy. *President condemns 'obsession' with economic growth*. 2023. URL: https://www.irishtimes.com/politics/2023/04/28/president-condemns-obsession-with-economic-growth/.

[585] R. Chaplin-Kramer et al. *Mapping the planet's critical natural assets*. 2023. URL: https://www.nature.com/articles/s41559-022-01934-5.

[586] United Nations. *System of Environmental Economic Accounting — seea.un.org*. URL: https://seea.un.org/content/homepage.

[587] United Nations. *Revision of the System of Environmental-Economic Accounting (SEEA)*. 2012. URL: https://unstats.un.org/unsd/statcom/43rd-session/documents/SEEA-Central-Framework-Ch1-E.pdf.

[588] Bram Edens and et al. Maes Joaquin. *Establishing the SEEA Ecosystem Accounting as a global standard*. 2022. URL: https://www.sciencedirect.com/science/article/abs/pii/S2212041622000092.

[589] United Nations. *2024 Global Assessment*. 2024. URL: https://seea.un.org/content/2024-global-assessment.

[590] Anati Prabhakar, Shalanda D. Young, and Gina M. Raimondo. *National Strategy to Develop Statistics for Environmental Decisions*. 2023. URL: https://bidenwhitehouse.archives.gov/wp-content/uploads/2023/01/Natural-Capital-Accounting-Strategy-final.pdf.

[591] *Wealth Accounting and the Valuation of Ecosystem Services*. 2020. URL: https://www.wavespartnership.org/about-us.

[592] New Bedford Fishing Heritage Center. *Industry Timeline*. URL: https://fishingheritagecenter.org/exhibits/industry-timeline/.

[593] David Tudor. *Landmark decision to reopen the Grand Banks commercial cod fishery*. 2024. URL: https://www.cmscoms.com/?p=39949.

[594] *Stock SMART - Status, Management, Assessments & Resource Trends*. 2021. URL: https://apps-st.fisheries.noaa.gov/stocksmart?stockname=Atlantic%5C%20cod%5C%20-%5C%20Gulf%5C%20of%5C%20Maine%5C&stockid=10508.

[595] *Atlantic Cod*. 2000. URL: https://www.fisheries.noaa.gov/species/atlantic-cod.

[596] Rahman Dr. S. et al. *A New National Strategy to Reflect Natural Assets on America's Balance Sheet OMB*. 2022. URL: https://www.whitehouse.gov/omb/briefing-room/2022/08/18/a-new-national-strategy-to-reflect-natural-assets-on-americas-balance-sheet/.

[597] K. Raworth. *What on Earth is the Doughnut?* 2017. URL: https://www.kateraworth.com/doughnut/.

[598] J. Bansard and M. Schröder. *The Sustainable Use of Natural Resources: The Governance Challenge. International Institute for Sustainable Development*. 2021. URL: https://www.iisd.org/articles/deep-dive/sustainable-use-natural-resources-governance-challenge.

[599] *Herbert Spencer*. 2019. URL: https://en.wikipedia.org/wiki/Herbert%5C_Spencer.

[600] *Microwaves: A Haven for Bacterial Diversity*. 2024. URL: https://tecscience.tec.mx/en/biotechnology/bacteria-in-microwave/.

[601] G. News. *Scientists Create Hybrid Robot Combining Fungus And Machine*. 2024. URL: https://ground.news/article/robot-controlled-by-a-mushroom-blends-living-organisms-and-machines%5C_209d8f?utm%5C_source=mobile-app%20%5C&%20utm%5C_medium=newsroom-share.

[602] L. Boissoneault. *What Really Turned the Sahara Desert From a Green Oasis Into a Wasteland?* 2017. URL: https://www.smithsonianmag.com/science-nature/what-really-turned-sahara-desert-green-oasis-wasteland-180962668/.

[603] *Global Assessment Report on Biodiversity and Ecosystem Services*. 2019. URL: https://www.ipbes.net/global-assessment.

[604] World Bank Group. *Water Security Financing Report 2024*. 2024. URL: https://www.worldbank.org/en/topic/water/publication/water-security-financing-report-2024.

[605] United Nations. *Goal 6, department of economic and social affairs*. 2024. URL: https://sdgs.un.org/goals/goal6.

[606] Save the Redwoods League. *Giant Sequoia*. 2025. URL: https://www.savetheredwoods.org/redwoods/giant-sequoias/.

[607] Save the Redwoods League. *Sonoma Coast Redwoods*. 2024. URL: https://www.savetheredwoods.org/.

[608] *Coast Redwoods, Save the Redwoods League*. 2014. URL: https://www.savetheredwoods.org/redwoods/coast-redwoods/.

[609] P. Tugwell. *Europe's $42 Billion Effort to Fight Wildfires Is an Uphill Battle*. 2024. URL: https://www.insurancejournal.com/news/international/2024/08/19/788859.htm.

[610] Sierra Club. *Forest Conservation*. 2025. URL: https://www.sierraclub.org/grassroots-network/forests-climate/forest-conservation.

[611] Singapore National Population Talen Devision. *How is the Total Fertility Rate calculated?* 2019. URL: https://www.population.gov.sg/media-centre/articles/how-is-the-tfr-calculated/.

[612] World Bank Group. *World Bank Open Data*. 2024. URL: https://data.worldbank.org/indicator/NY.GDP.PCAP.CD?locations=US.

[613] *Total fisheries and aquaculture production.* 2023. URL: https://openknowledge.fao.org/server/api/core/bitstreams/9df19f53-b931-4d04-acd3-58a71c6b1a5b/content/sofia/2022/world-fisheries-aquaculture-production.html.

[614] Food and Agriculture Organization of the United Nations. *Consumption of aquatic foods.* https://openknowledge.fao.org/server/api/core/bitstreams/9df19f53-b931-4d04-acd3-58a71c6b1a5b/content/sofia/2022/consumption-of-aquatic-foods.html. 2023.

[615] F. Richter. *This chart shows which countries consume the most or least fish.* 2022. URL: https://www.weforum.org/agenda/2022/11/chart-shows-countries-consume-fish-food-security/.

[616] I. Tiseo. *Annual Production of Plastics Worldwide from 1950 to 2020.* 2022. URL: https://www.statista.com/statistics/282732/global-production-of-plastics-since-1950/.

[617] S. Dutchen. *Microplastics are everywhere.* 2024. URL: https://magazine.hms.harvard.edu/articles/microplastics-everywhere.

[618] *Microplastics Found in Human Brains.* 2024. URL: https://e360.yale.edu/digest/microplastics-human-brains.

[619] *General Assembly Adopts Two Resolutions on Oceans, Highlighting Mounting Threats to Marine Resources, Need to Tackle Rising Sea Levels, Damage to Ecosystems.* 2023. URL: https://press.un.org/en/2023/ga12569.doc.htm.

[620] D. Berkelaar. *Terra Preta: the Secret of Rich, Dark Soil in the Amazon.* 2024. URL: https://www.echocommunity.org/resources/194f747a-ebc6-42b7-8a60-2c9cc990d7f0.

[621] J. Burton. *U. S. Geological Survey Releases 2022 List of Critical Minerals.* 2022. URL: https://www.usgs.gov/news/national-news-release/us-geological-survey-releases-2022-list-critical-minerals.

[622] J. Rachman. *US may block Indonesia nickel on forced labor issues.* 2024. URL: https://asiatimes.com/2024/09/us-may-block-indonesia-nickel-on-forced-labor-issues/.

[623] C. Louise. *How Long Does It Take For Leaves To Decompose?* 2023. URL: https://www.connect-green.com/understanding-natural-leaf-decomposition/.

[624] P. Jeandet et al. *The Role of Sugars in Plant Responses to Stress and Their Regulatory Function during Development.* 2022. URL: https://www.mdpi.com/1422-0067/23/9/5161.

[625] Q. Yang, D. Zhao, and Q. Liu. *Connections Between Amino Acid Metabolisms in Plants: Lysine as an Example*. 2020. URL: https://www.frontiersin.org/journals/plant-science/articles/10.3389/fpls.2020.00928/full.

[626] Perrin B. *Past eight years confirmed to be the eight warmest on record*. URL: https://wmo.int/news/media-centre/past-eight-years-confirmed-be-eight-warmest-record.

[627] UCAR Center for Science Education. *Predictions of Future Global Climate*. URL: https://scied.ucar.edu/learning-zone/climate-change-impacts/predictions-future-global-climate.

[628] Turner B. *Global carbon emissions reach new record high in 2024, with no end in sight, scientists say*. URL: https://www.livescience.com/planet-earth/climate-change/global-carbon-emissions-reach-new-record-high-in-2024-with-no-end-in-sight-scientists-say.

[629] Younger S. *Arctic Sea Ice Near Historic Low; Antarctic Ice Continues Decline*. URL: https://www.nasa.gov/earth/arctic-sea-ice-near-historic-low-antarctic-ice-continues-decline/.

[630] Stein T. and Allen M. *2024 Arctic Report Card documents rapid, dramatic change*. URL: https://www.climate.gov/news-features/understanding-climate/2024-arctic-report-card-documents-rapid-dramatic-change.

[631] Joe Fisher. *Natural disasters killed thousands around the world, caused billions in damage in 2024*. URL: https://www.upi.com/Top_News/World-News/2024/12/20/2024-weather-events-climate-disasters/4411733768309/.

[632] Catalin M. and Goldin H. *Drones, planes or UFOs? Americans abuzz over mysterious New Jersey sightings*. URL: https://apnews.com/article/drones-new-jersey-buzz-6ef6cfdd5dd6508da22b3c3a1d2f6718.

[633] Denton John W. H. *iccwbo.org*. URL: https://iccwbo.org/wp-content/uploads/sites/3/2024/11/2024-ICC-Oxera-The-economic-cost-of-extreme-weather-events.pdf.

[634] Anon. *The human cost of disasters: an overview of the last 20 years (2000-2019)*. URL: https://www.undrr.org/publication/human-cost-disasters-overview-last-20-years-2000-2019.

[635] Anon. *The War Economy of the Allies: Strategies and Impact*. URL: https://militarysaga.com/war-economy-of-the-allies/.

[636] Anon. *New report: extreme weather events cost economy $2 trillion over the last decade*. 2024. URL: https://iccwbo.org/news-publications/policies-reports/new-report-extreme-weather-events-cost-economy-2-trillion-over-the-last-decade/.

[637] Sayed Kazmi. *2024: A Year in Review – Global Weather Extremes*. 2024. URL: https://blog.worldweatheronline.com/weather/2024-a-year-in-review-global-weather-extremes/.

[638] Anon. *Fossil Fuel Subsidy Tracker*. 2024. URL: https://fossilfuelsubsidytracker.org/.

[639] Christian Thorsberg. *80 Percent of Global CO_2 Emissions Come From Just 57 Companies*. 2024. URL: https://www.smithsonianmag.com/smart-news/since-2016-80-percent-of-global-co2-emissions-come-from-just-57-companies-report-shows-180984118/.

Index

Thanks to...

Antoine Seronde, whose stimulating ideas inspired this book and whose thoughtful first reading pointed out errors and helped keep me on track.

First reader Larry Beck for encouraging me to continue when confidence lagged.

My loving wife, Laurie Beckelman, for her support and sharp editorial eye.

Fact checker Sara Hashimi for meticulous line-by-line fact checking that helped weed out misinterpretations and errors. But do blame me and not Sara for any false statements remaining in the text.

Typesetting gurus Peter Flynn, author of *Formatting Information: A beginner's introduction to typesetting with LaTeX*; Amy Hendricson, TeXnology Inc.; and Abdullah Sandhu for helping me navigate the challenges of LaTeX. I take full blame, however, for any and all less-than-elegant interior layout and formatting.

Sri Kartini, aka HAREYRA, for cover design.

Activists, researchers, and organizations worldwide working hard to remedy historic assaults on our natural world.

And...

Profound apologies to my grandchildren and all children of the world for the legacy of environmental peril we've bequeathed upon them.

Lloyd R. Prentice

Lloyd has an MA degree from Stanford University in Public Affairs Communication. His career spans corporate communication, academia, magazine publishing, and educational software development. His works include *Aya Takeo*, a three-volume manga for young adults; two novels, *Freein' Pancho* and *The Gospel of Ashes*; a technical programming book, *Build It with Nitrogen: the Fast-Off-the-Block Erlang Web Framework*; and more than 25 educational computer software products for major publishers.

Lloyd lives in a mixed eastern white pine and northern red oak forest four miles from the Atlantic Ocean in a town noted for saltwater marshes. He spends winter months in a small coastal villiage on the Gulf of California side of Baja California Sur, Mexico. He hopes that forests, oceans, marshes, and the scenic vistas of Baja provide future generations with as much joy as they give him.

www.ingramcontent.com/pod-product-compliance
Ingram Content Group UK Ltd.
Pitfield, Milton Keynes, MK11 3LW, UK
UKHW020657041026
15590UKWH00001B/326

9 780982 589274